Research Methods for Constructi

D1458579

arrer   Please return

234 011

Research Methods for Construction

Third Edition

Richard Fellows
Department of Real Estate and Construction
The University of Hong Kong

and

Anita Liu
Department of Civil and Building Engineering
Loughborough University

WILEY-BLACKWELL

A John Wiley & Sons, Ltd., Publication

NORWICH CITY COLLEGE			
Stock No.	234011		
Class	624.072 FEL		
Cat.	A2	Proc	JWL

This edition first published 2008
© 2008 by Blackwell Publishing Ltd

Blackwell Publishing was acquired by John Wiley & Sons in February 2007. Blackwell's publishing programme has been merged with Wiley's global Scientific, Technical, and Medical business to form Wiley-Blackwell.

Registered office
John Wiley & Sons Ltd, The Atrium, Southern Gate, Chichester, West Sussex, PO19 8SQ, United Kingdom
350 Main Street, Malden, MA 02148-5020, USA

Editorial office
9600 Garsington Road, Oxford, OX4 2DQ, United Kingdom

For details of our global editorial offices, for customer services and for information about how to apply for permission to reuse the copyright material in this book please see our website at www.wiley.com/wiley-blackwell.

The right of the author to be identified as the author of this work has been asserted in accordance with the Copyright, Designs and Patents Act 1988.

All rights reserved. No part of this publication may be reproduced, stored in a retrieval system, or transmitted, in any form or by any means, electronic, mechanical, photocopying, recording or otherwise, except as permitted by the UK Copyright, Designs and Patents Act 1988, without the prior permission of the publisher.

Wiley also publishes its books in a variety of electronic formats. Some content that appears in print may not be available in electronic books.

Designations used by companies to distinguish their products are often claimed as trademarks. All brand names and product names used in this book are trade names, service marks, trademarks or registered trademarks of their respective owners. The publisher is not associated with any product or vendor mentioned in this book. This publication is designed to provide accurate and authoritative information in regard to the subject matter covered. It is sold on the understanding that the publisher is not engaged in rendering professional services. If professional advice or other expert assistance is required, the services of a competent professional should be sought.

Library of Congress Cataloging-in-Publication Data

Fellows, Richard, 1948-
  Research methods for construction / Richard Fellows and Anita Liu. – 3rd ed.
    p. cm.
  Includes bibliographical references and index.
  ISBN-13: 978-1-4051-7790-0 (pbk. : alk. paper)
  ISBN-10: 1-4051-7790-X (pbk. : alk. paper) 1. Building–Research–Methodology. I. Liu, Anita. II. Title.

  TH213.5.F45 2008
  624.072–dc22

                                                                    2008002534

A catalogue record for this title is available from the British Library

Set in 10/13pt Palatino by Newgen Imaging Systems (P) Ltd, Chennai, India
Printed in Singapore by Utopia Press Pte Ltd

1 2008

Contents

Preface to Third Edition

We welcome the opportunity to both update and improve this book in order to extend and enhance its usefulness for researchers. The amendments reflect comments and suggestions from colleagues on the first two editions of the book as well as our own research progression into new areas of interest and study and the methods we have adopted and considered, together with feedback from students on undergraduate, postgraduate and research degrees whom we have supervised and to whom we have provided modules in research methods. We extend our thanks to all who have taken the trouble to advise us of potential improvements.

Throughout the production of this edition, as previously, we have endeavoured to be rigorous in scope and depth of content and to incorporate practical examples of applications of the methods to construction-oriented research. Our involvement with CIB (Conseil International du Bâtiment (International Council for Research and Innovation in Building and Construction)) groups, conferences and publications has proved invaluable in bringing us into contact with world-leading researchers in many construction fields which has greatly enriched our appreciation of the ever-growing wealth of research relating to construction both directly and indirectly.

However, we are also ever more aware of the pressures under which researchers must work, especially in terms of funding restrictions and quantity, as well as quality, of outputs – particularly, papers in refereed journals. At one end, information searching is vastly facilitated by the internet, whilst, at the other, journals proliferate. Both developments raise important issues of validity and reliability – the web remains (largely) unrefereed and unpoliced and so, the individual information-seeker must be the initial judge of reliability of what is retrieved whilst the proliferation of journals requires many more referees, and editors, to be 'gatekeepers' for maintenance and enhancement, of quality.

Despite the pressures and gripes, research is, for us, at least, the most stimulating and rewarding of pursuits, both in carrying out scholarship and empirical investigations ourselves and in assisting others as colleagues or supervisors. The sheer joy of valid discovery remains boundless; we hope this new volume helps all its readers to enjoy research too.

Anita Liu
Richard Fellows
Hong Kong, December 2007

Preface to Second Edition

'The scientist is not the person who knows a lot but rather the person who is not prepared to give up the search for truth.' Popper (1989, p. 334); reporting Marx and Engels.

A discipline or profession is established by developing a body of knowledge which is unique – that body of knowledge is produced through research. Construction draws on a wide variety of established subjects, including natural sciences, social sciences, engineering and management, and applies them to its particular context and requirements. Only by use of appropriate methodologies and methods of research, applied with rigour, can the body of knowledge for construction be established and advanced with confidence.

Although a number of texts are available discussing research methodologies and methods generally, there is a notable lack of such books in construction. Statistics, philosophy, natural and social sciences have produced relevant texts; this book is aimed at the broad discipline of construction. In particular, the contents of this book will be useful to students of building, civil engineering, architecture, construction management and all forms of surveying, whether researching for dissertations for Bachelors or Masters degrees or undertaking research for Masters degrees or Doctorates. Further, the book will be helpful to practitioners and students in these disciplines in providing guidance on how to instil rigour in problem-solving and on producing reports and publications.

The approach adopted in the book is to outline the process of research: the initial recognition that research is necessary; the development of a proposal; the execution of the research; the drawing of conclusions; and the production and presentation of the final report. The book comprises three main parts – producing a proposal, executing the research and reporting the results. The book discusses the main issues in research and examines the primary approaches – both qualitative and quantitative. The methods adopted for scientific and engineering experiments and simulations are evaluated as well as those employed for research into managerial issues, and social and economic investigations.

In considering the requirements for data and data analyses, the book presents discussion of important statistical considerations and techniques.

These enable the researcher to appreciate the issues which need to be evaluated in devising how research may be carried out effectively and efficiently in the practical environment of modern construction activity. Thus, the book considers a range of methodologies and methods to facilitate selection of the most appropriate research approach to adopt (from an informed perspective). It provides sufficient depth in examination of the subject materials to facilitate the execution of research projects.

Increasingly, pleas are voiced seeking special treatment (leniency) for research in construction, based on arguments concerning the particular nature of construction and the problems of research which ensue. Unless the research proposed and undertaken in construction can withstand scrutiny on the same bases as all other research, the discipline will fail to advance adequately. As construction is of major importance to all societies and economies, it is essential that the discipline advances as rapidly and as rigorously as possible.

This book results from the combined experiences of the authors in executing, supervising and managing many types of research projects over a number of years in the United Kingdom and Hong Kong in particular. This second edition has been produced to extend the scope of coverage, especially with respect to qualitative research. In this endeavour, we are grateful to the many colleagues in academia and beyond who have taken the trouble to provide valuable and constructive criticism of the first edition.

The production of this new edition has encouraged us to scrutinise the total content and to produce a volume that is more comprehensive, both in scope and critical comment on the methods discussed.

Once again we offer our deepest thanks to Julia Burden and her colleagues at Blackwell Science who have been so encouraging and patient with us throughout the preparation of the book. We apologise for fraying their nerves on occasion and assume full responsibility for the content, including any errors, omissions and contentious statements.

We hope that all readers will find the book stimulating and useful. Good luck in your research.

Richard Fellows
Anita Liu

Part 1

Producing a Proposal

Chapter 1

Introduction

The objectives of this chapter are to:

- introduce the concept of research;

- provide awareness of different *classifications of research*;

- outline the essentials of *theories and paradigms*;

- discuss the various *research styles*;

- introduce quantitative and qualitative approaches;

- consider *where*, and how, *to begin*.

The concept of research

Chambers English Dictionary defines research as:

- a careful search
- investigation
- systematic investigation towards increasing the sum of knowledge.

For many people, the prospect of embarking on a research project is a daunting one. However, especially for people who are associated with a project-oriented industry, such as property development, building design, construction, or facilities management, familiarity with the nature of projects and their management is a significant advantage. Dr Martin Barnes, an ex-chairperson of the Association of Project Managers (APM), has described a project as a task or an activity which has a beginning (start), a middle and an end that involves a process which leads to an output (product/solution). Despite the situation that much research is carried out as part of a long-term 'rolling' programme, each individual package of research is itself a project – an entity which is complete in itself, whilst contributing to the overall programme.

Indeed, any work which assists in the advancement of knowledge, whether of society, a group or an individual, involves research; it will involve enquiry and learning also.

Research: a careful search/investigation

Research can be considered to be a 'voyage of discovery', whether anything is discovered or not. In fact, it is highly unlikely that no discovery will result because discovery can concern the process of investigation as well as the 'technical subject' (the topic of investigation). Even if no new knowledge is apparent, the investigation may lend further support for extant theory. What is discovered depends on the question(s) which the research addresses, the patterns and techniques of searching, the location and subject material investigated and the analyses carried out. The knowledge and abilities of researchers and their associates are important in executing the investigative work and, perhaps more especially, in the production of results and the drawing of conclusions. Being open-minded and as objective as possible is vital for good research.

Research: contribution to knowledge

The Economic and Social Research Council (ESRC) defines research as '... any form of disciplined inquiry that aims to contribute to a body of knowledge or theory' (ESRC, 2007). That definition demonstrates that the inquiry must be appropriately deigned and structured and that it is the intent of the inquiry which is important (to distinguish from casual inquiries) rather than the outcome *per se*.

The *Concise Oxford Dictionary* (1995) provides a more extensive definition of research as 'the systematic investigation into and study of materials, sources etc. in order to establish facts and reach new conclusions'. Here the emphasis lies on determining facts in order to reach new conclusions – hence, new knowledge. The issue of 'facts' is not as clear, philosophically speaking, as is commonly assumed, and will be considered later.

The dictionary continues: 'an endeavour to discover new or collate old facts etc. by the scientific study of a subject or by a course of critical investigation'. Here there is added emphasis on the method(s) of study; the importance of being scientific and critical is reinforced.

Therefore, research concerns *what* (facts and conclusions) and *how* (scientific; critical) components.

Traditionally, the essential feature of research for a doctoral degree (PhD) is that the work makes an original (incremental) contribution to knowledge. This is a requirement for a PhD, and many other research projects also make original contributions to knowledge. A vast number of research projects synthesise and analyse existing theory, ideas, and findings of other research, in seeking to answer a particular question or to provide new insights. Such research is often referred to as scholarship; scholarship forms a vital underpinning for almost every type of research project.

Despite its image, research is not an activity which is limited to academics, scientists etc.; it is carried out by everyone many times each day. Some research projects are larger, need more resources and are more important than others.

Example
Consider what you would do in response to being asked, 'What is the time, please?'
   Having understood the question, your response process might be:

- look at watch/clock
- read time
- formulate answer
- state answer ('The time is . . .').

In providing an answer to the original question, a certain amount of research has been done.

A learning process

Research is a learning process . . . perhaps the only learning process.

Commonly, teaching is believed to be the passing on of knowledge, via instructions given by the teacher, to the learner. Learning is the process of acquiring knowledge and understanding. Thus, teaching exists only through the presence of learning and constitutes a communication process to stimulate learning; teaching is 'facilitation of learning'. If someone is determined not to learn, they cannot be forced to do so, although they may be persuaded to learn through forceful means.

Contextual factors affecting research

Research does not occur in a vacuum. Research projects take place in contexts – of the researcher's interests, expertise and experiences; of human contacts; of the physical environment etc. Thus, despite the best intentions and vigorous precautions, it seems inevitable that circumstances, purpose etc., will impact on the work and its results. The fact that research is being carried out will itself influence the results, as described in the Hawthorne Investigations of Elton Mayo (1949) and noted in the writings of Karl Popper (1989) on the philosophy of research. Research is never a completely closed system. Indeed, much (good) research is, of necessity, an *open* system which allows for adaptability.

As research is *always* executed in context, it is important to consider the contextual factors, the *environmental variables*, which may influence the results through their impacting on the data recorded. Such environmental variables merit consideration in tandem with the *subject variables* – dependent, independent and intervening (see Fig. 1.1) – of the topic of study. The choice of methodology/methodologies is important in assisting identification of all relevant variables, their mechanisms and amounts of impact.

Example

Consider Boyle's Law. Boyle's Law states that, at a constant temperature, the volume of a given quantity of a gas is inversely proportional to the pressure upon the gas, i.e.

$$V \propto \frac{1}{P}$$

$$PV = constant$$

Laboratory experiments to examine Boyle's Law attempt to measure the volumes of a particular quantity of gas at different pressures of the gas. The temperature is the environmental variable, to be held constant, the pressure is the independent variable and the volume is the dependent variable (following the statement of Boyle's Law). The researcher's breathing on the equipment which contains the gas may alter the temperature (otherwise constant) slightly and it will influence the results, though possibly not enough to be recorded. In such cases, the uncontrolled effects of environmental variables which impact on the results so that the relationship found is not in strict compliance with the statement of Boyle's Law, are denoted 'experimental error'.

Boyle's Law, like the other gas laws, strictly applies only to a perfect gas, but for many 'practical' purposes, all gases conform to Boyle's Law. For this reason, the purpose of the research is likely to be an important determinant of how the experiment is performed and to what level of accuracy. Considerations, such as those noted in respect of Boyle's Law experiments, lead to research being classified as pure research and applied research. Slightly different views classify studies as either research or development whilst the

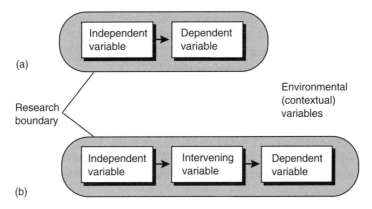

Fig. 1.1   'Causality chain' between variables (see also Fig. 4.1, p. 105).

purpose of a study often leads to academics' work being classified as research or consultancy.

Classifications of research

Pure and applied research

Frequently, classification of work is difficult, not only due to the use of 'fuzzy' definitions but, more importantly, because the work occurs within a continuum. At one end there is 'pure' or 'blue sky' research such as the discovery of theories, laws of nature etc., whilst at the other, applied research is directed to end uses and practical applications. Most academics are encouraged to undertake research towards the 'pure' end of the spectrum whilst practitioners/industrialists tend to pursue development work and applications. Of course, particularly in contexts like construction, the vast majority of research is a combination of 'pure' and 'applied' research – of theory and applications.

Essentially, development and applications cannot exist without the basic, pure research whilst pure research is unlikely to be of great benefit to society without development and applications. Unfortunately, much snobbery exists within the research and development sectors – those who work in one sector all too often decry (or fail to value) the contributions of others who work in different sectors. Fortunately, the advances of Japanese industry and many individual organisations which recognise and value the synergetic contributions of the various sectors of the research spectrum are fostering a change in attitude such that research and development activities are recognised as being different and complementary – each with particular strengths, approaches and contributions to make.

Often, the difference concerns the questions to be addressed rather than the approaches adopted. Pure research is undertaken to develop knowledge, to contribute to the body of theory which exists – to aid the search for the 'truth'. Applied research seeks to address issues of applications: to help solve a practical problem (the addition to knowledge is more 'incidental' than being the main purpose). The (not always material) distinction may be articulated as being that pure research develops scientific knowledge and so asks 'is it true?' whilst applied research uses scientific knowledge and so asks 'does it work?'

Commonly, research, especially applied research (located towards the developmental end of the research spectrum), involves solving problems. A simple dichotomous classification of types of problem is:

(1) *Closed* (ended) problems – simple problems each with a correct solution. The existence of the problem, its nature and the variables involved can be identified easily. Such problems are common, even routine, and so can be dealt with easily (often via heuristics/routines) to give the single correct solution. The problems are 'tame'.
(2) *Open* (ended) problems – tend to be complex; the existence of the problem may be difficult to identify, the situation is likely to be dynamic and so, the variables are difficult to isolate. Finding a solution is hard and may require novel ideas (e.g. through 'brainstorming'). It may not be (very) evident when a solution has been reached and many alternative solutions are likely to be possible. Such problems are 'wicked', 'vicious' or 'fuzzy' and may well concern/involve insight.

Clearly, most problems requiring research for their solution are likely to be open ended. However, in solving problems there are many sources of influence (bias) which may impact on the people involved – not least the approaches adopted for solving and the solutions determined for closed-ended problems.

Quantitative and qualitative research

The other primary classification system concerns the research methods adopted – broadly, quantitative and qualitative research. Quantitative approaches adopt 'scientific method' in which initial study of theory and liter-ature yields precise aims and objectives with proposition(s) and hypotheses to be tested – conjecture and refutation may be adopted, as discussed by authors such as Popper (1989). In qualitative research, an exploration of the subject is undertaken without prior formulations – the object is to gain understanding and collect information and data such that theories will emerge. Thus, qualita-tive research is a precursor to quantitative research. In an 'advanced' body of knowledge, where many theories have been developed and laws have been established, quantitative studies of their applicabilities can be undertaken without the need to determine theories and such afresh, thereby avoiding, 'reinventing the wheel' for each new study.

Generally, quantitative approaches provide 'snapshots' and so, are used to address questions such as what, how much, how many? Thus, the data, and results, are instantaneous or cross-sectional (e.g. compressive strength of a concrete cube; number of firms in an industry; market price of an item; content of an Architect's Instruction). Qualitative approaches seek to find out why things happen as they do; to determine the meanings which people attribute to events, processes and structures, etc. Many qualitative studies use data regarding people's perceptions to investigate aspects of their social world; others seek to 'go deeper' to address people's assumptions, prejudices, etc. to determine their impacts on behaviour and, thence, (organisational/project) performance.

The fundamental issues in designing any research, and so, underpin the selection of quantitative, qualitative or combination approaches, concern the research question and constraints and, perhaps most particularly, what is to be measured and the requirements of reliability and validity.

Sometimes qualitative research is assumed to be an easy option, perhaps in an attempt to avoid statistical analyses by persons who do not excel in mathematical techniques. Such an assumption is seriously flawed – to execute a worthwhile research project using qualitative methods can be more intellectually demanding than if quantitative methods had been employed. The use of qualitative methodologies should not necessarily be assumed to be a 'soft option'.

Irrespective of the nature of the study, rigour and objectivity are paramount throughout. Drenth (1998, p. 13) defines objectivity as '. . . the degree to which different observers of judges are able to record the data in the same manner. Judgement or classification of data in scientific research should not be substantially influenced by the subjectivity of the observer'. Thus, it is helpful if all the researchers agree the definitions of terms, metrics for collecting the data and the related protocols. Commonly, qualitative data, which are subjective data (such as obtained in opinion surveys), can and should be analysed objectively, often using quantitative techniques. However, one should not lose sight of the richness which qualitative data can provide and, often, quantitative data cannot. Triangulation – the use of qualitative and quantitative techniques together to study the topic – can be very powerful to gain insights and results, to assist in making inferences and in drawing conclusions, as illustrated in Fig. 1.2.

Research requires a systematic approach by the researcher, irrespective of what is investigated and the methods adopted. Careful and thorough planning are essential and, especially where large amounts of data are collected,

Fig. 1.2   Triangulation of quantitative and qualitative data.

rigorous record keeping is vital – in the study of theory and previous work (literature) as well as in the field work.

The impact of the researcher must be considered, both as an observer, experimenter, etc., whose presence may impact on the data collected and the results derived, and also through bias which may be introduced in data collection, analyses and inferences. Such biases may be introduced knowingly – to examine the subject from a particular viewpoint – or unknowingly, perhaps by asking 'leading questions'.

> *Example*
> Consider the question, 'Do you not agree that universities are under-funded?'
>   The phrasing, 'Do you not agree that . . .', suggests that the respondent ought to agree that universities are under-funded and so, asking such a question is likely to yield more responses of agreement than if the questions were phrased more objectively.
>   The question could be phrased much more objectively, 'Do you believe that universities are:
>
> (1) funded generously, or
> (2) funded adequately, or
> (3) funded inadequately?'

Even phrasing the question in that way, although removing the 'agreement bias' is incomplete as it assumes that all the respondents have a belief about the topic – some may not and so, a fourth possibility of 'no opinion' is appropriate. Unfortunately, that additional possibility also allows respondents to opt out of expressing their opinion!

Other categories of research

Further categorisation of types of research accords with the purpose of the research (question) as set out below.

- *Instrumental* – to construct/calibrate research instruments, whether physical measuring equipment or as tests/data collection (e.g. questionnaires; rating scales). In such situations the construction etc. of the instrument is a technological exercise; it is the evaluation of the instrument and data measurement in terms of meaning which renders the activity scientific research. The evaluation will be based on theory.
- *Descriptive* – to systematically identify and record (all the elements of) a phenomenon, process or system. Such identification and recording will be done from a particular perspective and, often, for a specified purpose; however, it should always be done as objectively (accurately) and as comprehensively as possible (this is important for later analysis). The research may be undertaken as a survey (possibly of the population identified) or as case study work. Commonly, such research is carried out to enable the subject matter to be categorised.
- *Exploratory* – to test, or explore, aspects of theory. A central feature is the use of hypotheses. Either an hypothesis is set up and then tested via research (data collection, analyses, interpretation of results) or a complex array of variables is identified and hypotheses are produced to be tested by further research.
- *Explanatory* – to answer a particular question or explain a specific issue/phenomenon. As in exploratory studies, hypotheses are used but here, as the situation is known better (or is defined more clearly), theory etc. can be used to develop the hypotheses which the research will test. Also, this could be a follow-on from exploratory research which has produced hypotheses for testing.
- *Interpretive* – to fit findings/experience to a theoretical framework or model; such research is necessary when empirical testing cannot be done (perhaps due to some unique aspects – as in a particular event of recent history, e.g. 'the Asian financial crisis of 1997'). The models used may be heuristic (using rules of thumb) – in which variables are grouped

according to (assumed) relationships – or ontological, which endeavour to replicate/simulate the 'reality' as closely as possible.

A further categorisation of research concerns what is being investigated – product, process or both. Research in construction includes all three categories; research into structural integrity is product oriented (e.g. strength properties of materials etc.), construction management research tends to be process oriented (e.g. organisational culture of construction firms) or both process and product (e.g. the impact of different procurement approaches on project and project management performance). Van de Ven (1992) identifies a process as '... a sequence of events that describes how things change over time'.

Theories and paradigms

Losee (1993, p. 6) depicts Aristotle's inductive–deductive method for the development of knowledge as shown in Fig. 1.3. He notes that, 'scientific explanation thus is a transition from knowledge of a fact [point (1) in the diagram] to knowledge of the reasons for the fact [point (3)]'.

Development of knowledge

Popper (1972, 1989) argues that scientific knowledge is different from other types of knowledge because it is falsifiable rather than verifiable, tests can only corroborate or falsify a theory, the theory can never be proved to be true. No matter how many tests have yielded results which support or corroborate a theory, results of a single test are sufficient (provided the test is valid) to

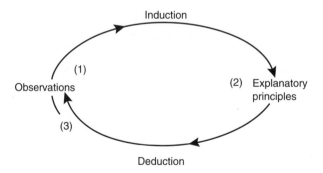

Fig. 1.3   Aristotle's inductive–deductive method (*source*: Losee 1993).

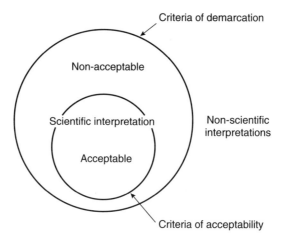

Fig. 1.4 Depiction of the approach to the advancement of knowledge, as advocated by Galileo (*source*: Losee 1993).

falsify the theory – to demonstrate that it is not always true. The more general application for acceptability in scientific investigation is shown in Fig. 1.4.

Different philosophies consider that scientific theories arise in diverse ways. Cartesians, who hold a 'rationalist' or 'intellectual' view, believe that people can develop explanatory theories of science purely through reasoning, without reference or recourse to the observations yielded by experience or experimentation. Empiricists, maintain that such pure reasoning is inadequate, it is essential to use experience from observation and experimentation to determine the validity or falsity of a scientific theory. Kant (1934) noted that the scope of peoples' knowledge is limited to the area of their possible experience; speculative reason beyond that, such as attempts to construct a metaphysical system through reasoning alone, has no justification.

Nagel (1986) suggests that the scientist adopts a 'view from nowhere' which implies the possibility of total objectivity and that phenomena exist totally independently of any observer. Conversely, Kuhn (1996) notes that 'what a man sees depends both upon what he looks at and also upon what his previous visual–conceptual experience has taught him to see'.

Tauber (1997) observes that, as science has evolved, so the notion of what constitutes objectivity has changed such that different branches of science require/employ different standards of 'proof'.

Dialectic, a development of 'trial and error', can be traced back to Plato who employed the method of developing theories to explain natural phenomena and followed this by a critical discussion and questioning of those theories; notably whether the theories could account for the empirical observations adequately. Thus, commonly, scientists offer theories as tentative solutions to problems; the theory is criticised from a variety of perspectives; testing the theory occurs, by subjecting vulnerable or criticised aspects of the theory to the most severe tests possible. The dialectic approach, following Hegel and discussed by authors such as Rosen (1982), is that a theory develops through the dialectic triad – thesis, antithesis and synthesis. The theory advanced initially is the thesis; often, it will provoke opposition and will contain weak points which will become the focus of opposition to it. Next, the opponents will produce their own counter-theory, the antithesis. Debate will continue until recognition of the strengths and weaknesses of the thesis and antithesis are acknowledged and the strengths of each are conjoined into a new theory, the synthesis. This is likely to regenerate the cycle of dialectic triad.

History, of course, has a role to play as it is likely to be influential, especially qualitatively, on how people think and behave in developing, criticising and interpreting theories. Popper (1989) uses the term 'historicism', whilst Clegg (1992) employs 'indexicality' to consider history's impact on how people understand, interpret and behave. Indexicality is a person's understanding etc. of terms and is determined by that person's background, socialisation, education, training etc. Marx's broad view was that the development of ideas cannot be understood fully without consideration of the historical context, notably the conditions and situations of their originator(s). It is possible to explain both formal social institutions (such as the UK parliament, the Sorbonne, the Supreme Court of USA, the Tokyo stock exchange, or the Royal Institution of Chartered Surveyors) and informal social institutions (such as friendship groups), by examining how people have developed them over the years.

Testing a theory

A *theory* is a system of ideas for explaining something; the exposition of the principles of science. Bacharach (1989) provides an amplified definition '. . . a theory may be viewed as a system of constructs and variables in which the constructs are related to each other by propositions and the variables are related to each other by hypotheses. The while system is bounded by the theorist's assumptions . . .'. Constructs are 'terms which, though not observational either directly or indirectly, may be applied or even defined on the basis of observables' (Kaplan 1964, p. 55). A variable as an observable entity which may assume two or more values (Schwab 1980).

Popper (1972) notes four approaches to testing a theory:

- 'The logical comparison of the conclusions among themselves, by which the internal consistency of the system is tested.

- The investigation of the logical form of the theory, with the object of determining whether it has the character of an empirical or scientific theory.
- The comparison with other theories, chiefly with the aim of determining whether the theory would contribute a scientific advance should it survive our various tests.
- The testing of the theory by way of empirical applications of the conclusions which can be derived from it.'

In particular, science provides rules for how to formulate, test (corroborate/falsify) and use theories.

Boolean logic states that concepts are polar in nature – they are either true or false. However, scientific theories are not of that form; they are not well defined, and so it is appropriate to consider a theory as being accepted due to the weight of supporting evidence (until falsified); rather akin to fuzzy logic. The value or usefulness of a theory may not be demonstrated by the use of probability alone; such probability must be considered in conjunction with the information contained in the theory. Broadly based, general theories may be highly probable but vague, due to their low information content; whilst precise or exact theories, with a high information content, may be of much lower probability. Theories with a high information content tend to be much more useful, which leads Blockley (1980) to require that appropriate measures to corroborate theories should be designed such that only theories with a high information content can achieve high levels of corroboration.

Tests (empiricism) can only corroborate or falsify a theory, as noted by Lakatos (1977). Losee (1993, p. 193) outlines Hempel's (1965) notion of three stages for evaluating a scientific hypothesis:

'(1) Accumulating observation reports which state the results of observations or experiments;
(2) Ascertaining whether these observations confirm, disconfirm or are neutral toward the hypothesis; and
(3) Deciding whether to accept, reject or suspend judgement on the hypothesis in the light of this confirming or disconfirming evidence.'

Husserl (1970, p. 189) asserts that 'the point is not to secure objectivity but to understand it'.

Scientific theories must be testable empirically. If a theory is true and one fact is known, often, another can be deduced. For example: if a theory states 'all clay is brown' and a sample provided is known to be clay, the deduction is that the sample will be brown. Provided the general statement of the theory is correct, in this case that all clay is brown, the deductive reasoning to go from the general statement to the specific statement, that the sample of clay is brown, is valid. However, discovery of clay which is a colour other than brown will falsify the general theory and so, require it to be modified, if not abandoned. Hence, deduction is 'safe', given corroboration of the theory/hypothesis, but it does not allow knowledge to be advanced.

Inductive reasoning – from the specific example to the general statement – is not valid. A *hypothesis* is a supposition/proposition made, as a starting point for further investigation, from known facts. (However, in formal research terms, a *proposition* concerns *constructs* and relationships between them whilst a *hypothesis* concerns *variables* and relationships between those – see Chapter 5.) Induction is useful to yield hypotheses, such as that by inspecting a variety of samples it may be hypothesised that all clay is brown. Thus, whilst the hypothesis remains corroborated rather than falsified, deductions can be made from it. Advances are made by use of induction. As knowledge advances, hypotheses may require qualifying statements to be appended to them – such as that all clay of a certain type and found in a given location, is brown – such auxiliary statements lend precision by raising the information content of the hypothesis or theory.

Thus, deductive reasoning occurs within the boundaries of existing knowledge (and may reinforce those boundaries), whilst inductive reasoning is valuable in extending or overcoming boundaries to current knowledge but should be employed with due caution – scientifically, through the use of hypotheses to be tested. Thus, Orton (1997) notes that 'Deductive research rhetorics tend to proceed from theory to data (theory, method, data, findings), while inductive research rhetorics tend to proceed from data to theory (method, data, findings, theory)'.

Exceptions to established general principles are called *anomalies* – instances in which the theory fails to provide a correct prediction of the particular reality. The presence of an anomaly usually promotes re-examination of the general principles/theory and, following further detailed investigation and use of the dialectic triad (see p. 14), the modification of the theory so that all the known instances are incorporated correctly.

The *fallacy of affirmation* occurs when certain observations apparently lead to particular conclusions regarding further relationships which appear to follow from the observations. However, without investigation of the validity of those conclusions on the basis of logical theory and empirical observation, false and misleading conclusions may ensue.

For example: Fact (1) Some penguins are flightless birds
               Fact (2) Some penguins are chocolate biscuits

False conclusion: Some flightless birds are chocolate biscuits

A paradigm

A *paradigm* is a theoretical framework which includes a system by which people view events (a lens). The importance of paradigms is that they operate to determine not only what views are adopted, but also the approach to questioning and discovery. Hence, much work concerns verification of what is expected or/and explanation of unexpected results to accord with the adopted, current paradigms. As progressive investigations produce increasing numbers and types of results which cannot be explained by the existing paradigms' theoretical frameworks, paradigms are modified or, in more extreme instances, discarded and new ones adopted – the well-known 'paradigm shift'.

Normally, the advance of knowledge occurs by a succession of increments, hence it is described as evolutionary. Only rarely are discoveries made which are so major that a revolutionary advance occurs. Often, such revolutionary advances require a long time to be recognised and more time, still, for their adoption, such as Darwin's theory of evolution. Hence, in terms of scientific progress, a theory which is valid at a given time is one which has not been falsified, or one where the falsification has not been accepted. Whilst objectivity is sought, research does have both cultural and moral contents and so, a contextual perspective, especially for social science research, is important to appreciate the validity of the study.

Kuhn (1996, p. 37) asserts that '. . . one of the things a scientific community acquires with a paradigm is a criterion for choosing problems that . . . can be assumed to have solutions . . .. A paradigm can . . . insulate a community from those socially important problems that are not reducible to the puzzle form because they cannot be stated in terms of the conceptual and instrumental tools the paradigm supplies.'

Positivism

Positivism originates in the thinking of Auguste Comte (1798–1857). It recognises only non-metaphysical facts and observable phenomena, and so is closely related to rationalism, empiricism and objectivity. Positivism asserts, in common with one branch of the Cartesian duality, that there are observable facts which can be observed and measured by an observer, who remains uninfluenced by the observation and measurement. Clearly, there is a strong relation to quantitative approaches.

However, the presence of 'facts' independent of the observer, and the feasibility of totally objective and accurate observation are being increasingly challenged. Whilst certain facts are, indeed, likely to exist independently of observation, this may be relevant and true as regards the 'natural world' only – the natural laws of the universe. Inevitably, observation and measurement affect what is being observed and measured (such as the issues involved in experiments to measure the temperature of absolute zero). Further, the matters of what is to be observed and measured, by whom, how, when, etc. are all determined by human decisions. Measurement may not be accurate for a variety of reasons, such as parallax, instrument error, etc.

In apparently separating reality of the natural world from those who attempt to observe and measure it, scientific positivism maintains the Cartesian duality to (supposedly) yield consistency of perception – the same inputs under the same circumstances yield the same outputs/results – the principle of replication.

Thus, Chia (1994) contrasts positivist and Kantian approaches as 'Positivist theories . . . maintain that . . . laws and principles are empirically discoverable, while Kantian theory insists that the basic categories of logic, time and space are not "out there" but are inherent constituents of the mind.'

Interpretivism

The interpretive paradigm is particularly valuable for research in management (and other social arenas) by indicating that reality is constructed by the persons involved. Thus, one person's reality, derived by observations and perceptions and modified by socialisation (upbringing, education and training) is likely to be different from another's. Therefore, truth and reality are social constructs, rather than existing independently 'out there' and so, researchers should endeavour to determine truth and reality from the participants' collective perspective(s) – to see things through their eyes. Such determination is likely to require extensive discussion with the participants, in order to achieve agreement on the representation (description) of their truth and reality and subsequent, further discussion to verify that the researcher's representation is correct.

As the interpretive paradigm is more likely to feature in qualitative studies (although it is applicable to quantitative research), there also exists the risk of influence (bias) by powerful participants who may be either individuals or groups. Therefore, the impact of social structure should be considered, including the perspective of *structuralists*, who argue that structure is fundamental to how society operates and to the determination of its values, customs, etc. This may, of course, be 'interactive cycling' societal values help to determine social structure, which then impacts on values, and so on.

Knowledge, then, may be regarded as constituting reality with a human component in that it is what, perhaps only for a time and place, counts as reality in being accepted as such by individuals or the population. Science is a mechanism for establishing and refining knowledge, as noted above, but with a focus on validation – itself a human process.

Tauber (1997, p. 3) notes that 'science is indeed a social phenomenon, but a very special one, because of the constraints exerted by its object of study and its mode of analysis'.

Pickering's (1992, p. 1) view is that 'scientific knowledge itself had to be understood as a social product'. That perspective is echoed by Pettigrew (1997) who asserts that 'Actions are embedded in contexts which limit their information, insight and influence . . . the dual quality of agents and contexts must always be recognised. Contexts and shaping and shaped, Actors are producers and products . . . interchange between agents and contexts occurs over time and is cumulative.'

The objectivity requirement of scientific positivism requires that knowledge of the observer is excluded. If personal knowledge (Polanyi 1962) – including intuitions and insights – are actually excluded, questions arise as to how investigations are instigated, how they are carried out and how conclusions are formulated. If we assume that investigations – research projects – do not just happen by pure chance but are initiated by cognitive motivation (interest, career development), then decisions (human, goal-directed actions) are taken to answer the basic investigative questions. Further, such motivational drives are determined by society and are likely to reflect and to perpetuate current perspectives of proper investigation of subjects and methods, often by use of 'immunising strategy', with only incremental, evolutionary change. Revolutions require bold challenges (Kuhn 1996) – such as that of Galileo.

Golinski (1990, p. 502) notes that the choices made by scientists and their managers '. . . are constrained by their aims or interests and by the resources they select to advance them'.

Perhaps it is more useful that the most suitable approaches to investigation, including the various forms of testing, are applied with rigour so that knowledge advances by employing models of maximum usefulness – following the high information content approach advocated by Blockley (1980). Such advances of science accept the roles of all types of inputs and testing – indeed, give credit to the role of triangulated approaches to modelling, testing, theory construction and paradigm 'drift' (a progressive, iterative movement between paradigms).

Whilst it is common for techniques themselves to be regarded as being 'value free', the selection of techniques to be used is 'value laden', due to 'indexicality' (e.g. Clegg 1992) and associated factors. However, techniques are devised and developed by researchers, and so, encapsulate the values of those involved in formulating the techniques – leading to debate over the merits of alternative techniques and their applications. Such potential for biases continues throughout the modelling process, and indeed may be made explicit – as in adopting a particular theoretical position to build an economic model.

Orton (1997) expresses the philosophical question underpinning the positivisism – interpretivism debate as '... whether theories are discovered, implying the existence of an objective world, or generated, implying the existence of a socially constructed world'. Thus, Pettigrew (1997) observes that 'Scholars are not just scientists, they remain obstinately human beings and as such are carriers of assumptions, values and frames of reference which guide what they are capable of seeing and not seeing'.

Models and hypotheses

A primary use of theory is to facilitate prediction. Instances where theories fail to predict correctly are anomalies. However, if a number of serious anomalies occur, the theory is likely to be rejected in favour of one which is more appropriate: one which explains all the occurrences more accurately. Hypotheses lead to theories which may be modified by auxiliary statements which 'particularise' them. Eventually, they may be rejected in favour of another theory of wider, accurate predictive ability. During the period of modifications and potential substitutions of theories, the 'competing' theories may be the subject of much debate in which advantages and disadvantages of each are considered to yield hierarchies of the theories continuously.

Another great value of theories is to enable researchers to produce models which show how the variables of a theory are hypothesised to interact in a particular situation. Such modelling is very useful in clarifying research ideas and limitations and to give insights into what should be investigated and tested.

Research styles

In determining what is the most appropriate approach (methodology and method(s)) to adopt – the research design – the critical consideration is the logic that links the data collection and analysis to yield results, and, thence, conclusions, to the main research question being investigated. The main priority is to ensure that the research maximises the chance of realising its objectives. Therefore, the research design must take into account the research questions, determine what data are required, and how the data are to be analysed.

Bell (1993) suggests styles of research to be Action, Ethnographic, Surveys, Case Study and Experimental. Yin (1994) considers that there are five common research strategies in the social sciences: *surveys, experiments* (including quasi-experiments), *archival analysis, histories* and *case studies*. Unfortunately, definitions of such styles vary and so, at best, the boundaries between the styles are not well defined.

Each style may be used for explanatory or descriptive research. Yin (1994) suggests that determination of the most appropriate style to adopt depends on the type of research operation (what, how, why, etc.), the degree of control that the researcher can exercise over the variables involved and whether the focus of the research is on past or current events. Requirements of the major research styles are set out in Table 1.1 at the end of this section.

Action research

Generally, action research involves active participation by the researcher in the process under study, in order to identify, promote and evaluate problems and potential solutions. Inasmuch as action research is designed to suggest and test solutions to particular problems, it falls within the applied research category, whilst the process of detecting the problems and alternative courses of action may lie within the category of basic research. The consideration of quantitative v. qualitative categories may be equally useful.

Action research (Lewin 1946) is where the research actively and intentionally endeavours to effect a change in a (social) system. Knowledge is used to effect the change which then creates knowledge about the process of change and the consequences of change (as well as of the change itself).

In programmes of action research, the usual cycle of scientific research (problem definition – design – hypothesis – experiment – data collection – analysis – interpretation) is modified slightly, by purpose of the action rather than by theoretical bases, to become 'research question – diagnosis – plan – intervention – evaluation', the 'regulative cycle' proposed by van Strien (1975) (in Drenth *et al.*, 1998).

Liu (1997) notes that action research is a shared process different from a hypothetical–deductive type of research. Thus, it is necessarily highly context dependent and so, is neither standardised nor permanent as it is reliant on the project and the knowledge and subjectivity/perceptions of the persons involved. Action research is operationalised to address a problem or issue which has been subject to structuring from use of theory.

The process of action research includes problem formation, action hypotheses, implementation, interpretation and diagnostic cycles (Guffond & Leconte 1995).

Action research is complex; the observer (who should provide a systematic perspective, relatively objectively) is involved and has the main role of creating a field for discussion and interpretation of the process and products. As change/innovation is the subject matter of the research (and the processes continue in parallel), coordination and evaluation mechanisms are necessary which involve both the researcher and the participants.

In consequence of the nature, and objectives, of action research, Henry (2000) asserts that three primary requirements exist:

(1) 'A trust-based relationship ... built up beforehand ... accepted by all parties ...
(2) The researchers will have fully accepted the firm's or institution's objectives for innovation or change by having negotiated the extent to which they will be involved and their freedom as regards access to information and interpretation.
(3) A research and innovation project will be jointly drawn up, which must be open ended with regard to the problems to be explored, but very precise in terms of methodology ...'

Ethnographic research

The ethnographic ((scientific) study of races and cultures) approach demands less active 'intrusion' by the researcher and has its roots in anthropology. The researcher becomes part of the group under study and observes subjects' behaviours (participant observation), statements etc. to gain insights into what, how and why their patterns of behaviour occur. Determination of cultural factors such as value structures and beliefs may result, but the degree of influence caused by the presence of the researcher, and the existence of the research project, will be extremely difficult (if not impossible) to determine.

The empirical element of ethnography requires an initial period of questioning and discussion between the researcher and the respondent to facilitate the researchers' gaining understanding of the perspectives of the respondent. Such interaction involves the 'hermeneutic circle' of initial questioning and transformation as a result of that interaction, all of which is embedded in the subject tradition (paradigm) of the researcher. Thus, 'Any interpretive act is influenced, consciously or not, by the tradition to which the researcher belongs' (Baszanger & Dodier 1997).

A further consideration is how the researcher integrates the empirical data etc. into a holistic perspective. The researcher's expertise and experience of field investigations represents a crucial moment in his/her education, prior to which he may have accumulated dissociated knowledge that might never integrate into a holistic experience; only after this moment will this knowledge 'take definitive form and suddenly acquire a meaning that it previously lacked' (Levi-Strauss 1974, quoted in Baszanger & Dodier 1997).

Complementarily, a sociological or political perspective recognises that the investigator is part of the group being studied and so, is a viable member of the group and a participant in the group behaviour as well as being the observer – more akin to the action research approach.

Thus, the approach focuses attention on determining meanings and the processes through which the members of the group make the world meaningful to themselves and to others.

Surveys

Surveys operate on the basis of statistical sampling; only extremely rarely are full population surveys possible, practical or desirable. The principles of statistical sampling – to secure a representative sample – are employed for economy and speed. Commonly, samples are surveyed through questionnaires or interviews. Surveys vary from highly structured questionnaires to unstructured interviews. Irrespective of the form adopted, the subject matter of the study must be introduced to the respondents. For a given sample size of responses required, particular consideration must be given to the response rate (i.e. the percentage of subjects who respond) and number of responses obtained. Following determination of the sample size required, appropriate procedures must be followed to assist in securing the matching of responses to the sample selected. This is a special consideration for 'stratified' samples; samples classified into categories, usually by proportion of the total population under investigation or measured degrees of another important, continuous attribute.

Case studies

Case studies encourage in-depth investigation of particular instances within the research subject. The nature of the in-depth data collection may limit the number of studies, when research is subject to resource constraints. Case studies may be selected on the basis of their being representative – with similar requirements/conditions to those used in statistical sampling to achieve a representative sample, to demonstrate particular facets of the topic, or to show the spectrum of alternatives. (See also the detailed classification in Yin (1994).) Case study research may combine a variety of data collection methods, with the vehicle or medium of study being the particular case, manifestation or instance of the subject – such as a claim, a project, a batch of concrete.

Commonly, case studies employ interviews of key 'actors' in the subject of study; such interview data may be coupled with documentary data (such as in a study of a production process). Alternatively, a case study may be 'situational', such as a wage negotiation or determining safety policy, and for such research, several 'cases' may be studied by individual or combined methods of ethnography, action research, interviews, scrutiny of documentation etc. Hence, case studies constitute a distinct 'style' of research.

Case studies operate through theoretical generalisation, as for experiments, rather than empirical/statistical generalisation (as is the approach via surveys, which employ samples designed to be representative of the population).

Experiments

The experimental style of research is, perhaps, suited best to 'bounded' problems or issues in which the variables involved are known, or, at least, hypothesised with some confidence. The main stages in experimental design are shown in Fig. 1.5. Usually, experiments are carried out in laboratories to test relationships between identified variables; ideally, by holding all except one of the variables constant and examining the effect on the dependent variable of changing the one independent variable. Examples include testing the validity of Boyle's Law, Hooke's Law, and causes of rust experiments. However, in many cases, notably in social sciences and related subject fields, experiments are not conducted in specially built laboratories but in a dynamic social, industrial, economic, political arena. An example is Elton Mayo's 'Hawthorne Experiments' which took place in a 'live' electrical manufacturing company (Mayo 1949). Such instances are 'quasi-experiments' as the ability to control

Table 1.1   Requirements of different research styles/strategies.

Style/Strategy	Research Questions	Control Over Independent Variables	Focus on Events
Survey	Who, what, where, how many, how much?	Not required	Contemporary
Experiment/Quasi-experiment	How, why?	Required	Contemporary
Archival Analysis	Who, what, where, how many, how much?	Not required	Contemporary/past
History	How, why?	Not required	Past
Case Study	How, why?	Not required	Contemporary

Source: Derived from Yin 1994.

Fig. 1.5   Experimental design.

variables (independent and environmental) is limited and, often, coupled with measurement problems which impact on accuracy.

> *Example*
> Consider investigating client satisfaction with the provision of a construction project. What quantitative and what qualitative data are likely to be available readily on a case study of a construction project?
>
> Quantitative data would comprise time and cost performance derived from project records – predicted v. actual; quality might be considered from records of re-worked items, corrections required due to defects recorded during the maintenance period – measured by number, value etc.
>
> Qualitative data could present participants' perceptions of client satisfaction with respect to the performance criteria of cost, time and quality. Such data would be obtained through questioning of those participants' identification of the variables and hypothesising of their inter-relations. Research could proceed by endeavouring to hold all but one of the independent variables constant and examining the effects of controlled changes in the remaining independent variable on the dependent variable.
>
> In certain contexts, such as medical research, the sample under study may be divided into an experimental group and a control group. After the experimental period, the results from the groups may be compared to determine any differences between the groups' results which can be attributed to the experiment. In such cases, the members of the groups must not know to which group they belong; it is helpful also (to avoid possible bias in analysis), if those who carry out the analysis of results are not informed of which person is in each group either.
>
> Hence, experimentation is aimed at facilitating conclusions between cause and effect – the presence, extent etc. Experimentation is at the base of scientific, quantitative method.

Quantitative and qualitative approaches

It is quite common for small research projects to be carried out with insufficient regard to the array of approaches which may be adopted. This may be because the appropriate approach is obvious, or that resource constraints preclude evaluation of all viable alternatives, or it may be due to a lack of awareness of the alternatives. Such lack of awareness does not mean that the research cannot be executed well, but, often, it does mean that the work could have been done more easily and/or could have achieved more.

Usually, research methods and styles are not mutually exclusive although only one, or a small number of approaches, will normally be adopted due to resource constraints on the work. The different approaches focus on collection and analysis of data rather than examination of theory and literature. The methods of collecting data impact upon the analyses which may be executed and, hence, the results, conclusions, values, validity and reliability of the study.

Quantitative approaches

Quantitative approaches tend to relate to positivism and seek to gather factual data, to study relationships between facts and how such facts and relationships accord with theories and the findings of any research executed previously (literature). Scientific techniques are used to obtain measurements – quantified data. Analyses of the data yield quantified results and conclusions derived from evaluation of the results in the light of the theory and literature.

Qualitative approaches

Qualitative approaches seek to gain insights and to understand people's perceptions of 'the world' – whether as individuals or groups. In qualitative research, the beliefs, understandings, opinions, views etc. of people are investigated – the data gathered may be unstructured, at least in their 'raw' form, but will tend to be detailed, and hence 'rich' in content and scope. Consequently, the objectivity of qualitative data often is questioned, especially by people with a background in the scientific, quantitative, positivist tradition. Analyses of such data tend to be considerably more difficult than with quantitative data, often requiring a lot of filtering, sorting and other 'manipulations' to make them suitable for analytic techniques.

Analytic techniques for qualitative data may be highly laborious, involving transcribing interviews etc. and analysing the content of conversations. Clearly, a variety of external, environmental variables are likely to impact on the data and results and the researchers are likely to be intimately involved in all stages of the work in a more active way than usually is acceptable in quantitative studies.

Triangulated studies

Both qualitative and quantitative approaches may adopt common research styles – it is the nature and objectives of the work together with the (consequent) nature of the data collected and analytic techniques employed which determine whether the study may be classified as qualitative or quantitative. Given the opportunity, of course, triangulated studies may be undertaken. As triangulated studies employ two or more research techniques, qualitative and quantitative approaches may be employed to reduce or eliminate disadvantages of each individual approach whilst gaining the advantages of each, and of the combination – a multi-dimensional view of the subject, gained through synergy. Thus, triangulation may be used for entire studies (such as by investigating a topic from several, alternative paradigms or/and research methodologies) or for individual part(s) of a study (such as collecting quality performance data from archival records of defects, questionnaires administered to project participants, and results of participant observation). Jick (1979) notes that between methodology triangulation seeks to enhance a study's external validity whilst within methodology triangulation seeks to enhance internal validity and reliability.

Whatever approach, style or category of research is adopted, it is important that the validity and applicability of results and conclusions are appreciated and understood. In particular, it is useful to be demonstrably aware of the limitations of the research and of the results and conclusions drawn from it. Such limitations etc. are occasioned by various facets of the work – sampling, methods of collecting data, techniques of analysis – as well as the, perhaps more obvious, restrictions of time, money and other constraints imposed by the resources available. Hence, it is very helpful to consider the constraints, methods etc. at an early stage in the work to ensure that the best use is made of what is available. Indeed, it may well be preferable to carry out a reduced scope study thoroughly than a larger study superficially – both approaches have validity but achieve different things.

Thus, whilst triangulation employs plural methods, 'bridges' involves linking two or more analytic formats (research methods) to make them more mutually informative, whilst maintaining the distinct contributions and integrity of each independent approach/discipline. Therefore, 'bridges' uses plural methods to link aspects of different perspectives.

Data sources

As with any project, the planning phase is crucial and it is wise to evaluate what is being sought and the alternative approaches available as early as possible. Of course, re-evaluations may be necessary during the course of the work, in instances such as where the data required prove to be unavailable. As data are essential to research, it is useful to consider what data are required, and alternative sources and mechanisms for collection during the planning phase. Use of surrogate data (indirect measures of what is sought) may have to be used, especially where the topic of study is a sensitive one (e.g. cost, safety, pricing, corruption, labour relations).

Where researchers have good contacts with potential providers of data, use of those sources is likely to ease the data collection process. If trust and confidence have been established, it is likely to be easier to obtain data and it may be possible to obtain data which might not be available otherwise. Certainly, trust and confidence are important considerations in data collection – the more sensitive the data, the more trust in the researcher which is required by the provider.

Especially for obscure and complex processes, and sensitive/historical subjects, finding sources of data/respondents may be difficult. However, once an initial source has been found it may be possible to find others (successively/progressively) by information from that initial source (analogous to references from a paper or book). The 'snowball' approach concerns the discovery and investigation of different sources for a particular event whilst the tracer approach moves between sources relating to the development/operation of a process.

In undertaking research in construction management, Cherns and Bryant (1984) note that, 'A basis must exist between the researchers and the [respondent] system for negotiating a relationship which has something to offer to the [respondent] as well as to the researchers.'

'Access must provide for deep and continued penetration into the [respondent] system at the earliest possible stage of the [building] project, preferably before the decision to [proceed].'

Often, it is essential to ensure that the providers of data cannot be traced from the output of the research. Statements ensuring anonymity are helpful as are methods which demonstrate anonymity in the data collection, such as not requiring names and addresses of respondents. However, anonymity must work. It is hardly providing anonymity if one identifies respondents as A, B . . . N rigorously in the research report but thanks respondents by name in the acknowledgements section.

Confidentiality is a similar, ethical issue to anonymity: anonymity refers to persons and organisations whilst confidentiality relates to the data. The two issues are closely related such that confidentiality concerns neither revealing data to anyone nor using the data for purposes other than those for whom the respondents have given permission. Both confidentiality and anonymity are very important components of research ethics, the moral underpinnings of which dictate that the *express, informed consent of the respondents must be obtained and adhered to rigorously.*

Occasionally, respondents wish to scrutinise a report prior to its 'publication'. Whilst such provision is useful in building confidence over anonymity issues, and may assist in ensuring accuracy of data etc., it may be regarded as an opportunity for the respondents to comment on the research, and, possibly, to demand changes – perhaps, to remove portions with which they disagree or which they dislike. Such changes should be resisted, provided the research has been conducted properly (accuracy of data and results, compliance with anonymity and confidentiality, etc.), as they would distort the research report and, thereby, devalue the work.

For applied research, it is increasingly popular to form a *steering group* of the principal investigators, industrialists and practitioners. The steering group helps to form the strategy for the execution of the work and to monitor and guide the research during its execution. The objective is to ensure the combination of rigorous research with practical relevance. Of course, there are spin-off benefits of the researcher's enjoying easier access to data via the committed practitioners, and the practitioners' gaining knowledge and insight of issues/problems which are important to them.

Where to begin

Research methodology refers to the principles and procedures of logical thought processes which are applied to a scientific investigation. *Methods* concern the techniques which are available (for data collection, analysis, etc.) and those which are actually employed in a research project. Any management of a research project must address certain questions in making decisions over its execution. The questions involved are:

- What?
- Why?
- Where?
- When?
- How?
- Whom?
- How much?

It is these questions which the study of this book will assist in answering or, rather, provide some information to help to reach an answer. By addressing the issues explicitly and logically, noting requirements, constraints, and *assumptions*, the progress through research projects will smooth and ease progressively as expertise and experience develop.

Often, a researcher is able to select a supervisor or supervisors. In selecting a supervisor, three considerations apply – that person's subject experience and expertise, research experience and expertise and, perhaps the most important factor differentiating potential supervisors, the ability to relate to and communicate well with the researcher. The best research tends to be executed by people who get on well together as well as possessing complementary skills and expertise.

It is important to determine the scope of the work at the outset; the most common problem is for a researcher to greatly overestimate what is required of the work, what can be achieved and the amount of work that can be done. It is a good idea to consult an experienced supervisor or 'third party' to ensure that the programme and scope of the research intended is *realistic*.

Example

What? Concerns selection of the topic to be researched with consideration of the level of detail. It is useful to note the resources available and constraints so that an appropriate scope of study can be determined.

Why? May command a variety of answers, each of which applies individually but some of which may apply in combination. So, 'required for a degree', 'required by employer', 'interest', 'career development', and possibly many other reasons, may be advanced to say why research is being undertaken. However, why a particular research project is being carried out or proposed, apart from the reasons given already, may be due to its being topical or because the researcher has expertise in that subject and wishes to use that expertise to acquire and advance knowledge in that field.

Where? Obviously all research occurs somewhere – the host institution may be a university, as well as the various places at which individual research activities occur – libraries, data collection points, visits to experts etc. It may be useful to consider the amount of travel, both cost and time, as an input to the strategy for executing the research.

When? The timing of the research and time available to carry it out will usually be specified. It will be necessary to produce a timetable for the work by dividing the time available between the component activities. Often, there will be restrictions on the time for data collection – allow for holiday periods, very busy periods etc.; what sequences of activities are necessary and what are the alternatives? To what extent can the activities overlap? A common problem is to devote insufficient time to planning the work and to forget, or, at least, to under-estimate, the time necessary for data analyses, production of results and conclusions and for preparation of the report. All too often the only real focus is on fieldwork (data collection) – such enthusiasm is healthy but must be kept under control.

How? Is the issue of methodology and of methods. In some instances, the methodology is obvious – virtually 'given' – as in computational fluid dynamics. Commonly, a topic may be investigated in a variety of ways, individually or in combinations, so a choice must be made. The choice will be influenced by the purpose of the research, the subject paradigm, the expertise and experience

continued

Example   Continued
of the researcher and supervisor (if any), as well as practical considerations of resource and data availabilities.

Whom? Four main groups of people are involved in the execution of research – the researcher, the supervisor, the sample personnel, who provide the data or access to it, and others who can help, such as laboratory technical staff. Naturally, a research project is 'commissioned' by someone, for instance, a university, as a requirement of a course of study, an academic agency such as a research council or a commercial agency, perhaps a government body, company consultant practice etc.

How Much? This issue concerns the resources which can be used. Many resources, such as money, are fixed but people's time tends to be rather flexible – especially the time input by researchers themselves. No research project is really completed from the researcher's point of view as there is always a bit more which could or ought to be done. Hence, each report contains a list of recommendations for further research.

Summary

This chapter has introduced some main concepts of research. A definition of research has been provided and the contexts of undertaking research have been discussed. Various approaches to research have been examined – notably qualitative and quantitative – together with their combination through 'triangulation'. Styles of study have been considered and questions which research projects address have been noted. The ethical issues of confidentiality and anonymity have been discussed and the essential need for objectivity has been emphasised.

References

Bacharach, S.B. (1989) Organizational theories: some criteria for evaluation, *Academy of Management Review*, **14**(4), 496–515.

Baszanger, I. & Dodier, N. (1997) Ethnography: relating the part to the whole. In: *Qualitative Research: Theory, Method and Practice* (ed. D. Silverman), pp. 8–23, Sage, London.

Bell, J. (1993) *Doing your Research Project*, 2nd edn, Open University Press, Buckingham.

Blockley, D.J. (1980) *The Nature of Structural Design and Safety*, Ellis Horwood Ltd, Chichester.

Cherns, A. & Bryant, D.T. (1984) Studying the clients' role in construction management, *Construction Management and Economics*, **2**, 177–184.

Chia, R. (1994) The concept of decision: a deconstructive analysis, *Journal of Management Studies*, **31**(6), 781–806.

Clegg, S.R. (1992) Contracts cause conflicts. In: *Construction Conflict Management and Resolution* (eds P. Fenn & R. Gameson), pp. 128–144, E. & F.N. Spon, London.

Drenth, P.J.D. (1998) Research in work and organizational psychology: principles and methods. In: *Handbook of Work and Organizational Psychology*, Vol. 1, 2nd edn (eds P.J.D. Drenth, H. Thierry and C.J. de Wolff), pp. 11–46, Psychology Press, Hove.

ESRC (2007) Research Ethics Framework (REF), Swindon: Economic and Social Research Council, www.esrc.ac.uk/ESRCInfoCentre/Images/ESRC_Re_**Ethics**_Frame_tcm6-11291.pdf-, accessed 20 June 2007.

Golinski, J. (1990) The theory of practice and the practice of theory: sociological approaches in the history of science, *ISIS*, **81**, 492–505.

Guffond, L.E. & Leconte, G. (1995) Le 'disposif': un outil de mire en orme et de conduite du changement industrial, *Sociologie du Travail*, **3**, 435–457.

Hempel, C.G. (1965) *Aspects of Scientific Explanation and other Essays in the Philosophy of Science*, Free Press, New York.

Henry, E. (2000) Quality Management Standardisation in the French construction industry, singularities and internationalisation projects, *Construction Management & Economics*, **18**(6), 667–677.

Husserl, E. (1970) *The Crisis of European Sciences and Transcendental Phenomenology* (trans. D. Carr), Northwestern University Press, Evanston (first published in 1935).

Kant, I. (1934) *Critique of Pure Reason*, 2nd edn (trans. F.M. Muller), Macmillan, New York.

Kaplan, A. (1964) *The Conduct of Inquiry*, Chandler, San Francisco.

Kuhn, T.S. (1996) *The Structure of Scientific Revolutions*, 3rd edn, The University of Chicago Press, Chicago.

Jick, T.D. (1979) Mixing qualitative and quantitative methods: triangulation in action, *Administrative Science Quarterly*, **21**(4), 602–611.

Lakatos, I. (1977) *Proofs and Refutations*, Cambridge University Press, Cambridge.

Levi-Strauss, C. (1974) *Anthropologie Structurale*, p. 49, Plon, Paris.

Lewin, K. (1946) Action research and minority problems, *Journal of Social Issues*, **2**, 34–36.

Liu, M. (1997) *Fondements et Practiques de la Recherche-action*, Logiques sociales, L'Harmattan, Paris.

Losee, J. (1993) *A Historical Introduction to the Philosophy of Science*, 3rd edn, Opus, Oxford.

Mayo, E. (1949) *The Social Problems of an Industrial Civilisation*, Routledge & Kegan Paul, London.

Nagel, T. (1986) *The View from Nowhere*, Oxford University Press, Oxford.

Orton, J.D. (1997) From inductive to iterative grounded theory: zipping the gap between process theory and process data, *Scandinavian Journal of Management*, **13**(4), 419–438.

Pettigrew, A.M. (1997) What is processual analysis?, *Scandinavian Journal of Management*, **13**(4), 337–348.

Pickering, A. (ed.) (1992) *Science as Practice and Culture*, The University of Chicago Press, Chicago.

Polanyi, M. (1962) *Personal Knowledge: Towards a Post-critical Philosophy*, The University of Chicago Press, Chicago.

Popper, K.R. (1972) *The Logic of Scientific Discovery*, Hutchinson, Oxford.

Popper, K.R. (1989) *Conjectures and Refutations: The Growth of Scientific Knowledge*, Routledge & Kegan Paul, London.

Rosen, M. (1982) *Hegel's Dialectic and its Criticism*, Cambridge University Press, Cambridge.

Schwab, D.P. (1980) Construct validity in organizational behavior. In: *Research in Organizational Behavior* (eds B.M. Straw & L.L. Cummings), Vol. 2, pp. 3–43, JAI Press, Greenwich, CT.

Strien, P.J. van (1975) Naar een methodologie van het praktijk denken in de sociale wentenschappen, *Naderlands Tijdschrift voor de Psychologie*, **30**, 601–619, quoted in Drenth, P.J.D. (1998) Research in work and organizational psychology: principles and methods. In: *Handbook of Work and Organizational Psychology, 2nd edn Vol. 1: Interaction to Work and Organizational Psychology* (eds P.J.D. Drenth, H. Thierry & C.J. de Wolff), pp. 11–46, Psychology Press, Hove.

Tauber, A.I. (1997) *Science and the Quest for Reality*, New York University Press, New York.

The Concise Oxford Dictionary of Current English (9th edn) (1995) (ed. D. Thompson), Clarendon Press, Oxford.

Van de Ven, A.H. (1992) Suggestions for studying strategy process: a research note, *Strategic Management Journal*, **13**(S1), 169–188.

Yin, R.K. (1994) *Case Study Research: Design and Methods*, 2nd edn, Sage, Thousand Oaks, CA.

Chapter 2

Topic for Study

> The objectives of this chapter are to examine the processes of:
>
> - selecting a topic;
>
> - writing a research proposal.

Selection of a topic

Very often, the most difficult task for any researcher is to select a topic for study and then to refine that topic to produce a proposal which is viable. Generally, people set targets which are far too high in terms of both the extent of the research which is possible and the discoveries which are sought. It is surprising to most new researchers how little (in scope) can be achieved by a research project and, hence, the necessity to restrict the study in order that adequate depth and rigour of investigation of the topic can be undertaken.

Resources

An important aspect to evaluate is the quantities of resources which can be devoted to the study. Often, it is helpful to calculate the number of person-hours, days, weeks, months or years, which are available for the research. Given a fixed amount of time and the period within which the research must be completed, and taking account of any flexibilities, the amount of work which can be undertaken begins to be apparent. Usually, a report of the work is required, and that report must be produced within the time frame, so, the period required to produce the report reduces the time available for executing the study itself.

Many people consider that undertaking a research project is 2% inspiration and 98% perspiration – clearly, research is *not an easy option*. Research is hard work, but it is often the most rewarding form of study. The satisfaction and sense of achievement derived from a project completed well can be enormous; the efforts are well worthwhile and provide the researchers with expertise, experience and insights for future work. Especially in the early days of a project, enthusiasm is a great asset – it is a major contributor to overcoming difficulties which will almost inevitably arise. Determination is valuable for a researcher as it will help to ensure that the project is seen through to completion.

Even in cases where a topic is given – such as where a researcher applies for a post to carry out a particular project advertised – there is selection of the topic by the prospective researcher. Generally, where a research project is part of a course of study, the choice of topic to research is made by the individual *but* that choice should not be made in isolation. Potential tutors, supervisors and mentors should be consulted, together with colleagues and, if possible, practitioners to assist in selecting a topic which is interesting, viable and appropriate to the context and the people concerned – most especially the *researcher*. The amount of time and effort spent in selecting and refining a topic and then planning to yield a proposal may appear very long, if not excessive. Invariably, it will be time well spent. Such formative stages are of paramount importance and, often, will be the main factor determining whether the research is a success.

Some academics believe that the Pareto distribution applies to research study. A Pareto distribution is the '80-20 rule' (Fig. 2.1); a small proportion of components have the major effect on the outcome. Applied to a construction project, about 20% by number of the components account for about 80% of the project cost. The Pareto distribution is believed to apply far and wide; it applies to programmes of study in that 80% of the work is completed

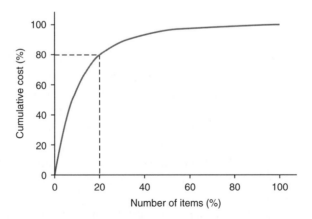

Fig. 2.1   Pareto distribution applied to the cost of a construction project.

(or becomes visible) in the last 20% of the time available – this is partly due to preparatory work being carried out in the early part of the study and so not being accessible, but also because certain people do not do the work until the deadline looms, due to other pressures or lack of programming. Requesting an extension of time may not be viewed favourably; indeed, if such a Pareto distribution does apply, it may be preferable not to grant an extension.

Subject selection

In most cases, researchers must confront the issue of subject selection. Although the selection will not be made from all subject areas but will be confined within the boundaries of particular disciplines, the possibilities are vast. Therefore, it is helpful to consider the process of subject selection as one of progressive narrowing and refinement – the degree to which those are taken being determined by the nature of the discipline and the appropriate research methods. Essentially, the process is strategic – the particular initial question is *what*?

In deciding any course of action, it is a good idea to undertake a SWOT analysis in which PEST factors are scrutinised. SWOT analysis requires determination of Strengths, Weaknesses, Opportunities and Threats. Strengths and Weaknesses are features of the individual or organisation and are *internal* factors, whilst Opportunities and Threats are present in the environment, so they are external, or exogenous, factors. PEST factors are the Political (and legal), Economic, Social and Technical forces in the environment. Consideration of the PEST factors assists analysis of the Opportunities and Threats. For examples see Fellows *et al.* (2002) and Newcombe *et al.* (1994). A sound strategy is to build on strengths and overcome weaknesses in seeking to take advantage of opportunities and to minimise the possible effects of threats.

In selecting a subject for a research project, it is useful to begin by constructing several lists:

List 1: Topics of interest to the researcher.
List 2: Personal strengths and weaknesses.
List 3: Topics of current interest in practice.
List 4: Data required for each topic.
List 5: Sources of data for each topic.
List 6: Research limitations for each topic.

The first list may contain topics of interest in quite broad terms. The second list should be of personal strengths and weaknesses. Certain strengths and weaknesses could influence the choice of topic directly whilst others might do so by consideration of the methodology and the nature of the analyses which would be appropriate. However, whilst building on strengths is good, an individual, interest-driven research project may be an excellent vehicle to extend the researcher's knowledge and experience. So, there are good arguments for pursuing a topic of research which is of particular interest to the researcher – it could serve to enhance career opportunities.

A third list might concern topics of current interest – what is being debated in the technical press – the 'sexy' or 'hot' topics of the day. Adopting such a topic for study should ease data collection owing to the amount of interest in the subject. However, many other researchers may select those topics for study also with the consequence that so many people are trying to collect similar data from a limited number of sources (often, organisations and their personnel) that the field may become 'saturated'. A second consideration is that interest in topics tends to be quite short-lived and the clever thing to do is to predict what the next issue of debate will be. In the UK construction industry, hot topics seem to have a 'life' of less than two years, although many 'cycle round' again and again.

> *Example*
> **List 1**   Begin by listing topics of interest in quite broad terms e.g. Concrete pre-stressing; Project procurement approaches; Price setting mechanisms. Such topics are too broad and insufficiently specific for certain types of research, but topics such as 'The culture of design and build contractors' or 'Effect of water content on bearing capacity of London Clay' could lend themselves to investigation.
>
> **List 2**   A second list of personal strengths and weaknesses may contain self-evaluations such as 'good at economics but quite weak in maths and statistics' or 'experienced in new build and in contracting but no experience in refurbishment or design consultancy'.
>
> **List 3**   Over recent years, 'hot topics' have included; buildability, sick building syndrome, life cycle costing, risk management, value management, quality assurance, quality management, refurbishment, facilities management, dispute resolution procedures, public–private partnerships. Other topics, although never getting quite 'red hot', may retain enduring interest; CAD, communications, information flows, payments mechanisms, inter-party relationships, price determination, environmental issues, client satisfaction, demolition of prestressed tower blocks.

Further lists should be compiled for each topic under consideration and should, by addressing the research question possibilities and the variables involved, determine the likely data requirements, sources of the data and access to data via industry or practice contacts. An employer may wish for particular issues to be investigated. Such projects should facilitate co-operation in data provision but care is needed over possible attempts to impose restrictions on publication of the contents of the report etc. Other lists may include the interests and expertise of potential supervisors and collaborators, and other researchers' work to which access can be gained readily.

Finally, it is likely to be helpful to list constraints and the resources available – these lists will be invaluable in determining the practicality of a proposed study and assist in determining what can be done, rather than trust what it would be 'nice' to do.

Example

List 4   Data requirements are determined by the topic of investigation, the research 'question(s)' and the variables involved – especially the independent and dependent variables. What is to be measured must, first, be defined so that appropriate metrics can be selected; that is especially important for qualitative data, such as quality of a construction project or component. However, quantitative data must be defined also – such as the capital cost of a project (to whom, when to measure, units of measurement, etc.).

List 5   Once the data requirements have been identified, the possible sources of data should be noted; some sources are likely to be better than others regarding ease of collecting the data, quality of the data (accuracy, amount of data), etc. Access to data should, ideally, match the data required, but is likely to be modified by practicalities of obtaining the data; especially if the data are 'sensitive' (e.g. costs, safety, corruption). Employers and sponsoring bodies as well as professional institutions may be helpful in securing access to the data needed.

List 6   Research limitations will be: the constraints on the resources available to execute the research and, hence, the scope of the study; data available; the methods/techniques employed. The time and resources available for the research, especially for dissertations and academic theses, are likely to be well-known, so it is the particular limitations which relate to the research which should be noted – especially data availability. However, if limitations should have been foreseen from the nature of the topic – sensitivity, the literature, experiences of other researchers etc. – care must be taken to explain why such limitations were not taken into account and avoided in the research design. In any event, it is important to note the reasons for what was done to overcome the limitations and to ensure validity of the resultant research.

It is best if the lists contain 'raw' ideas and do not contain the results of sophisticated evaluations – they come later. Essentially, the lists result from 'brainstorming' so it may be helpful to do the brainstorming in a group designed in such a way that people's ideas 'spark' further ideas in others – no evaluations or disparaging comments, just throw out the ideas, note them and think about further possible topics until no more emerge. Do not force things; if it is difficult to get a first idea, it may be helpful to visit the library, take a book at random, open it and pick a word at random and see if a research topic can be devised from that word. Such a serendipity approach may yield a novel and exciting topic (if horse races were predictable, everyone should win but would that be exciting?).

Choosing a topic

Having produced a number of lists of possible topics, constraints, require-ments etc., the next step is to begin some evaluation (judging possibilities against criteria/desires). Again, it is important to let the research topics be the driving force and the requirements and constraints be the parametric fac-tors, denoting the limits and extent of what may be done – they should not be the dictating factors even though they may 'loom large' from a student researcher's perspective.

Evaluating alternatives

Within overall subject domains, a number of topics will have emerged. As a background to the evaluation process, it may be helpful to consider two issues:

- 'What does the research seek to achieve?'
- 'What does the research seek to find out?'

In terms of achievement, apart from the obvious, important, but also mun-dane issue of satisfying the imposed purpose of carrying out the project, such as obtaining a good grade to contribute to a degree, the personal advancement aspects are important. Broadly, the achievement may be to extend understand-ing in a particular subject area, to broaden understanding in a new subject area by enhancing the depth of knowledge of the researcher or to gain knowledge in subjects where the researcher has less expertise. Of course, combinations of achievement in both ways may be possible, but the underlying issue is largely one of self-motivation, to cope with the 98% perspiration (hard work).

The second issue – what the research sets out to find out – concerns selecting a particular topic within the overall subject area decided already. So, the list of topics allocated to the overall subject area selected will be evaluated. Again, the lists of personal factors, data availability, etc. will be used in the evaluation, but retaining the approach of the topic as driver and the various constraints as parameters. It may be difficult to make a single choice, but whether one, two or three topics are the result of the evaluation, it will be helpful to consult literature – leading journals and reputable texts – to gain some preliminary, but further, understanding of the issues relating to the topics, whether they have been investigated extensively already and any likely problems. Secondly, it is useful to examine what particular terms are used, what the variables are, what data are needed and how those data may be collected.

A researcher should not be dismayed if the topic or closely related ones have been investigated previously. Much research, including scientific method, uses not only conjecture and refutation but, as an integral part of that philosophy, looks to replication of research. Studies of safety or performance in the UK construction industry are worthy of replication in Hong Kong, Malaysia, USA, China etc., partly to evaluate the methodology in other contexts, partly to re-examine the findings of the original study and partly to examine the situation, both absolutely and comparatively, between the locations of the studies. Indeed, replication, even if on a reduced scale, of a leading study, but in a different context, thereby changing only one variable if possible, has the great advantage of using a tested and substantiated methodology. This allows more emphasis to be placed on examination of results, whilst monitoring the methodology and providing any necessary critique due to the changed circumstances.

Therefore, evaluations of the topics from the perspectives of desires and constraints, often in discussion with colleagues, practitioners and the potential supervisor, will result in the selection of a single topic. The result of such an evaluation does not mean that details of the topic will not change. Research, by its nature, is a 'voyage of discovery', an iterative learning process, and investigation of the unknown is likely to require changes from *a priori* expectations. If it were not so, it might not be research, although the extent of changes should not lead to radical changes in the topic.

Refining a topic

The process of selection will continue for some time as investigations proceed and the topic (through considerations of definitions, variables and their relationships, aspects of theory, findings of previous work etc.) emerges and undergoes progressive refinement. The goal is to reach a state in which the topic is delineated sufficiently well for the aim and objectives of the research to be identified, and appropriate methodologies considered, to enable a research plan to be formulated, including a draft timetable.

One vital aspect is for the researcher to examine the 'ideal' data set from a research perspective and then to review it in the context of the practicalities

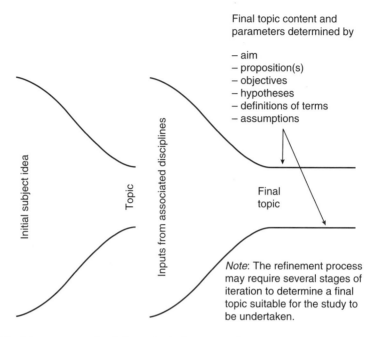

Final topic content and
parameters determined by

– aim
– proposition(s)
– objectives
– hypotheses
– definitions of terms
– assumptions

Initial subject idea

Topic

Inputs from associated disciplines

Final
topic

Note: The refinement process
may require several stages of
iteration to determine a final
topic suitable for the study to
be undertaken.

Fig. 2.2   The process of refining a topic for research.

of data collection (including alternative data from different sources, given access potential). Here, the researcher must have foresight and a perspective on practicality – *if you cannot get the data, you cannot do the research*.

The process of refining a topic for research is depicted in Fig. 2.2.

Writing the proposal

The outcome of the initial considerations and investigations will be a proposal for the research. The UK's Engineering and Physical Sciences Research Council (EPSRC) prescribes the format for proposals, as do most funding agencies. It comprises a form concerning the support requested and an outline of the project plus a six page 'case for support' outlining the project proposal in more detail.

Normally, for a degree of Bachelor or Master dissertation, a proposal of around four sides of A4 is adequate; a proposal for MPhil or PhD will be more extensive, but all proposals should be concise. Depending on the nature of the research proposed, the proposal should contain:

- Title
- Aim

- Proposition (if appropriate)
- Objectives
- Hypothesis (if appropriate)
- Methodology
- Programme
- List of primary references.

Usually, it is helpful to append a diagram showing how the variables envisaged are relevant to the research and how they are hypothesised to relate to each other. Often, that forms the basis of very helpful analytic discussions of the topic and aids in-depth thinking and investigation throughout. Cronbach and Meehl (1955) developed the concept of a *nomological network* to be a diagram of the theoretical framework (variables, constructs and relationships between them), the empirical framework (to indicate what is to be measured) and the linkages between the two frameworks.

Aim

The aim of a research project is a statement of what the research will attempt to do – often in the form of what is to be investigated, which is more appropriate for qualitative research, or what impact the main independent variables are believed to have on the dependent variable, an approach which is more suitable for quantitative studies. The aim is really a statement at the strategic level so, usefully, can be considered to be what the researcher would like to do if resource constraints and other constraints did not exist. Clearly, constraints do exist, and so the research should not be judged, once completed, against the yardstick of the statement of aim. Rather, the aim provides the identification of the context of what is attempted.

> *Example*
> To investigate the 'maintenance path' for local authority school buildings in UK through establishing maintenance needs and work execution mechanisms, and provide maintenance information to designers in an environment of resource constraints.

Proposition

A proposition is a statement, derived from the aim of a research project. Thus, it concerns the strategic level of the research and so, identifies the primary constructs and their (supposed) relationships. A proposition reflects the onto-logical (specific account of existence; a specification of a conceptualisation of concepts and relationships between them) and epistemological (theories of the sources, nature and limits of knowledge) views of the researcher which constitute the paradigm within which the research will be carried out. As any proposition relates to constructs, it must be refined further to yield oper-ational statements (as objectives and hypothesis) which comprise variables that can be measured. Thus, the form of statement constituting a proposition is very similar to a hypothesis (see below) but a proposition is at a higher level of abstraction; support (or otherwise) for a proposition is usually determined from testing of its constituent hypothesis (or hypotheses).

Objectives

The objectives are statements within the strategic statement of aim; they are statements at the tactical/operational level. Objectives take the aim of the research and, given the constraints, translate the aim into coherent, operational statements. These are statements which relate to each other logically but which are, each, self-sufficient also and describe what the research hopes to achieve or discover through the study.

The objectives specify what it is hoped will be discovered by the research – what will be known at completion of the project which was not known at its start *and* has been revealed by the research. However, for qualitative studies, the objectives may concern how the study will be undertaken and some details of what is to be studied. Both approaches are valid but they are different. It is important for researchers to recognise the differences and the consequences for the research resulting from the adoption of the alternative approaches. So, say, for a research project to investigate the impact of the culture of participants on the performance of construction projects, it may be useful to conduct a dual investigation – a sociologically based study providing a more qualitative approach in parallel with a construction management based study giving a more quantitative approach. Results of the individual investigations could, perhaps, be triangulated to yield synergy.

For ethnographic-type research, a broad statement of aim may be all that can be produced. It may be that in some instances, even that may not be viable – only identification of the subject area may be possible. The researchers will have notions of what is to be investigated but, due to lack of theory, prior investigation etc., they may not be able to produce detailed objectives and hence, methodology/methods. The approach is to collect all possible data rigorously and to use those data to structure the study from the relationships and patterns which emerge (a form of constructivism). It is probable that all knowledge began to be discovered from this approach – perhaps most obviously in the social sciences where people's behaviours had to be studied to detect patterns etc. from which hypotheses were generated for testing and the subsequent derivation of theories and 'laws'.

For most research projects, especially smaller ones, it is good discipline to restrict the project to a *single* aim and the objectives to about *three*. Such restriction promotes rigour in considering what the research is about and what can be achieved realistically. Keep the statements simple, especially for objectives, working with one independent variable in each which impacts on the dependent variable of the study, and ensuring that those variables can be identified.

> *Example*
> (1) To investigate any linkages between construction types and maintenance requirements.
> (2) To examine any relationship between age of buildings and their maintenance needs.
> (3) To determine the factors which impact on maintenance work execution for UK local authority school buildings.
> (4) To develop and test a model for maintenance of UK local authority school buildings.

Hypotheses

Certain studies, such as ethnographic research projects, may achieve the formulation of an hypothesis as their end result. Other studies may develop hypotheses as the result of initial investigations, whilst a third category will formulate hypotheses in the early stages of the work – as in experimental research projects. Usefully, a hypothesis can be regarded as the 'focus' of a research project – the statement of supposition of what are the important variables relating to the topic and how they relate to each other.

It is useful to note the inter-relationship between aim, objectives, proposition(s) and hypothesis. As the aim is the strategic statement of what the research seeks to discover, it leads to the development of a proposition(s). The objectives comprise operationally oriented refinements of the aim and, analogously, the hypothesis is derived from the proposition as it concerns the main variables (derived from the constructs in the proposition). Hence, the aim and objectives, and proposition and hypothesis should be developed in parallel, as illustrated in Fig. 2.3.

Irrespective of when a hypothesis is formulated, it is a statement of conjecture – it suggests a relationship between an independent and

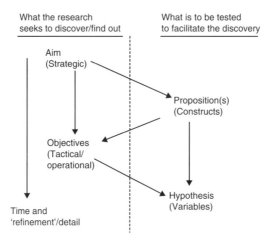

Fig. 2.3 Development of aim and objectives, and proposition and hypothesis.

a dependent variable and the nature of that relationship. The statement concerns direction in the relationship, known as causality.

Example

Consider the following hypothesis:

'The method of programming construction projects employed by contractors influences project performance, and hence participants' satisfaction with those projects.'

Apart from criticisms of the English and phrasing of the hypothesis, it contains two dependent variables – project performance and participants' satisfaction. This raises issues of what may be said about support for the hypothesis if, after testing, one part is supported and the other is not. Clearly, it would be preferable to split the hypothesis into two, or even three:

- programming–performance
- performance–satisfaction
- programming–satisfaction.

The performance–satisfaction relationship is implied in the hypothesis. To retain this in the study, it could be determined from theory and previous work but, for rigour and completeness, it should form an element of the research. A further complication is that participants is plural and so, whilst in any particular case some participants may be satisfied, it is unlikely that there will be many cases in which all participants are either satisfied or dis-satisfied. Who the participants are must be clarified. An added complication is what satisfaction is and how to measure it due to its subjective and individualistic nature.

Hypotheses are statements which are produced to be tested as objectively as possible. It is useful to have *one* main hypothesis, derived from the aim of the research, and sub-hypotheses relating to the objectives (if appropriate). Further sub-hypotheses may be included provided that they assist or clarify the research; too many will tend to promote confusion. Provided they are appropriate, hypotheses are extremely valuable in lending direction, constraints of relevance and objectivity to a research project. The goal must be to *test* the hypotheses objectively through data collection and analyses; usually, hypotheses are derived from theory and literature.

A common misconception is that once hypotheses have been established, the goal of the research is to prove them. That is neither possible nor desirable. Even the most extensive research will not prove anything absolutely, although it may establish a likelihood with an extremely high level of confidence, as is common in medical research. Seeking to prove or to disprove a hypothesis is likely to introduce bias into the research – contrary to the requirement of objectivity. So, if a researcher believes the hypothesis to be true, it may be a good idea to propose the opposite hypothesis to attempt to counter any bias in the researcher's initial beliefs, as there may be an innate tendency for researchers to seek to support the hypothesis of the study unwittingly.

Hypotheses also focus the work on relevant aspects and help to identify boundaries of the study in experimental and quantitative researches. Especially where resources are very limited, it is invaluable to be able to identify the boundaries for the study to ensure that effort is expended only where it will be relevant to the particular topic. Such an approach is alien to much qualitative work which seeks immersion in the subject matter to collect all possible data for analysis to see what, if anything, emerges.

Use of hypotheses indicates data requirements and suggests analyses to be performed. Naturally, other relationships may emerge and so, it is important to retain openness to new ideas etc. Although tests should examine what has been hypothesised, there must always be a preparedness and opportunity for other findings to emerge – comprehension rather than restriction must apply.

Methodology

Methodology, the principles of the methods by which research can be carried out, lies at the heart of research. Many good ideas remain uninvestigated and/or unfunded because the methodology (and the methods) has not been considered adequately. Many of the following sections are concerned with the major aspects of research methodologies and the detailed considerations of these have been left until later. It is vital that the methodology is given careful consideration at the outset of the research so that the most suitable approaches and research methods are adopted.

The importance of a good theoretical base cannot be over-estimated. That concept applies not only to the topic of research but to methodology and methods as well. Thus, Pettigrew (1997) notes that 'The practice of research is best informed by the theory of method which clarifies and makes explicit the range of guiding assumptions shaping the conduct of that research.'

Whenever possible, it is useful to draw a diagram of the variables likely to be involved and the hypothesised relationships between them. Attention should be given to:

(1) Definitions of the main terms involved, especially where terms have varied definitions.
(2) Note the assumptions which are made and the justifications for them.
(3) Critically review the theories, principles and literature relating to the subject matter of the research.
(4) Evaluate what analyses may be carried out with respect to data available, the objectives and any hypothesis.

For instance, soft systems methodology concerns social problems for tackling soft and ill-structured problems, such as value management, from the sociological perspective. It emphasises learning, human content, epistemologies and system models.

In Checkland's (1981, 1996) soft systems methodology (SSM), the perception of a problem situation is recorded in the first two stages of analysis, stage 1 of unstructured problem situation and stage 2 of problem expression. Stage 3 (root definitions) and stage 4 (conceptual model) then utilise systems ideas to conceptualise/predicate certain selected features of the problem situation. Based on the conceptual model, the prediction can then be compared with the realities of the problem (stage 5) to improve the problem situation in the last two stages of the methodology: changes to be made (stage 6) and implementation of action (stage 7). Since the desirable and feasible changes may create a new problem situation in the cyclic process, soft systems methodology is a learning process for accommodating/integrating conflicting interests among participants.

Example
Liu and Leung's (2002) paper develops a conceptual model of value management in construction procurement within the framework of goal-setting theory and considers that value management is aimed at the economic and managerial aspects of project development to focus the definitions of project goals which are obtained through the interaction between the client and the project team members. (See also later results Leung & Liu 2003.)

Programme

Milton Friedman, the Chicago School economist, stated that, 'Only surprises matter' (Friedman 1977, p. 12); perhaps that statement is more appropriate to research than to any other activity.

The production of a timetable or programme for any research project is essential to ensure the project's viability. It helps the researcher, and others, to decide how the activities should fit together and the time to be devoted to each one. Thus, once formulated and agreed, it will provide a basic yardstick against which progress may be monitored. However, due to the nature of research, it is important that the programme, and those using it, have sufficient flexibility to enable novel, potentially productive, lines of investigation to be noted or pursued in some way.

Normally, a simple bar chart showing the main research activities is adequate as a programme for the work. For smaller, quantitative projects, times for 'producing the report', 'finalising the proposal', 'theory and literature review' and 'data analysis' may be predicted with some confidence; programming these activities will indicate the amount of time and timing of the data collection. Despite overlaps of activities, the time available for the field work and analysis is likely to be quite constrained, so efficiency and effectiveness of data collection are essential.

Deliverables and industrial or practitioner support

Increasingly, applied research is required to focus on the provision of practically useful deliverables. On many fronts, research is being pushed, via the funding agencies, towards serving the needs of industry and commerce. Fundamental and qualitative research is unlikely to produce immediate, industrially useful deliverables, such as results and findings, directly although, in the long term, results of such work can be extremely important and far-reaching.

In Britain and many other western market-developed economies, industry and commerce tend to have short-term views and so, will support research projects only if they appear likely to produce profit-enhancing results quickly. This may, perhaps, be due to performance imperatives imposed by financiers. Such a view is likely to present less problems for short duration, applied research than for long duration, fundamental studies. However, support there must be, if only to facilitate collection of data. In seeking support, it is helpful to show the relevance of the work and to contact a person likely to have interest in the study and to be in sympathy with it. The support of an industrial or professional association will help. The offer of the provision of a summary of findings is not only good manners in showing appreciation of assistance and it can be given confidentiality, if necessary. Do ensure that such summaries *are provided* to the participants. Letters of request from the researcher's institution lend credibility to the work, ensure legitimacy of the study and should provide appropriate and sympathetic 'control' over sensitive facts.

Summary

This chapter has considered the processes by which a suitable topic for research, such as a dissertation, may be selected. Often, the selection process is one of narrowing from a subject area to a particular issue for study. Criteria and parameters must be evaluated along with the rationale for undertaking the research. Throughout, the requirement is for objectivity, especially in the formulation and subsequent testing of hypotheses.

References

Checkland, P. (1981) *Systems Thinking, Systems Practice*, pp. 59–191 and 299–319, John Wiley & Sons, USA.

Checkland, P. (1996) Soft systems methodology. In: *Rational Analysis for A Problematic World* (ed. J. Rosenhead), ch. 4, pp. 71–100, John Wiley & Sons, England.

Cronbach, L. & Meehl, P. (1955) Construct validity in psychological tests, *Psychological Bulletin*, **52**(4), 281–302.

Fellows, R.F., Langford, D.A., Newcombe, R. & Urry, S.A. (2002) *Construction Management in Practice*, 2nd edn, Blackwell Science, Oxford.

Friedman, M. (1977) *Inflation and Unemployment*, p. 12, Institute of Economic Affairs, London.

Leung, M.Y. & Liu, A.M.M. (2003) Analysis of value project goal specificity in value management, *Construction Management and Economics*, **21**(1), 11–19.

Liu, A.M.M. & Leung, M.Y. (2002) Developing a soft value management model. *International Journal of Project Management*, **20**(5), 341–349

Newcombe, R., Langford, D.A. & Fellows, R.F. (1994) *Construction Management 1: Organisation Systems*, Mitchell, London.

Pettigrew, A.M. (1997) What is processual analysis?, *Scandinavian Journal of Management*, **13**(4), 337–348.

Part 2
Executing the Research

Chapter 3

Initial Research

The objectives of this chapter are to:

- introduce the *research process*;

- discuss the requirements for the *initial research* and methods of reviewing theory and literature systematically and critically;

- stress the importance of assembling the theoretical framework;
  - consider the *philosophy and methodology* of the research;
  - examine the role of a theoretical model and constructs;

- emphasise the imperative of *proper referencing*.

The research process

In the 1980s, the Science and Engineering Research Council (SERC), the fore-runner of the Engineering and Physical Sciences Research Council (EPSRC) in UK, held a Specially Promoted Programme (SPP) in Construction Management and issued the following diagram (Fig. 3.1) of their view of the research process relating to the SPP. Initial studies provide the foundation for all the research work that follows.

Research is a dynamic process. Therefore, it must be flexible – implying, although not requiring, that a contingency approach will be helpful. Early in the study, links between problems (which may be either topics or issues), theories, previous findings and methods will be postulated. The links should form a coherent chain, and so may need to be adapted as the work develops and findings emerge. The goal must be to maintain coherence and complementarity; only by such an approach will the results and conclusions be robust. The research path, outlined in Fig. 3.2, is embodied in the research design, data collection and data analyses, encompassing both the nature of the data

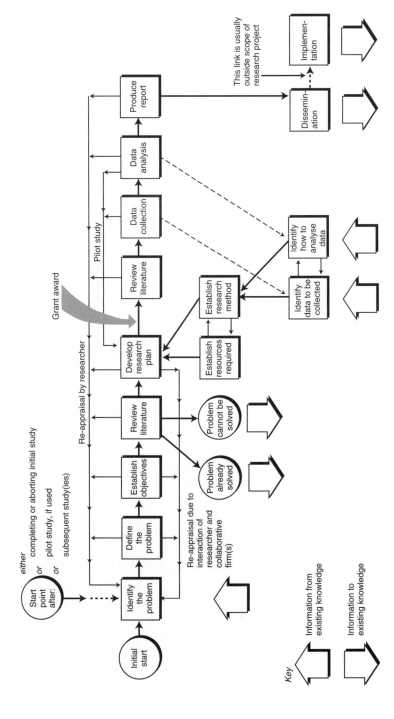

Fig. 3.1   Suggested construction management research process (*source*: SERC 1982).

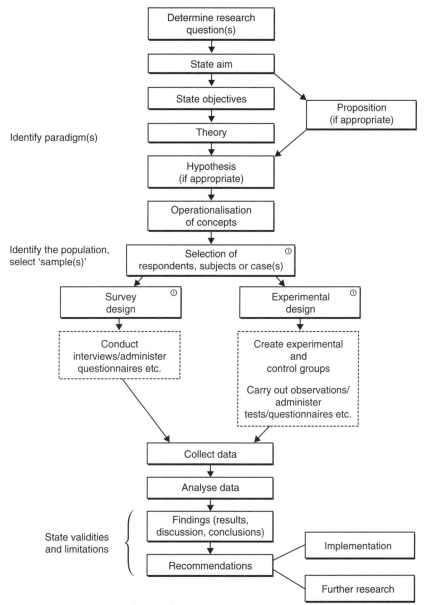

Note: ① It may be advisable to carry out these activities in the reverse sequence to that shown: identify the population and possible 'sample(s)' and select and design the survey(s) or experiment(s) then select the 'sample(s)' to be used (from theoretical and practical considerations).

Fig. 3.2   The research process (after Bryman & Cramer 1994).

and the methods used. Whatever method is adopted, it is essential that the research be conducted rigorously – that it is an objective, reliable and valid study.

Initial phase

Depending on the nature of the study (see Fig. 3.1), the initial work provides either the means for determining or confirming the aim, objectives and hypothesis or for confirming the topic for study. In either case, initial studies are essential to ensure that the research intended has not been carried out already and, more especially, to determine what has been researched already and what issues remain or have emerged for investigation. This helps to avoid repeating work which has been carried out already (unless the objective is to replicate a study for examining its reliability/validity) and to avoid making the same mistakes that previous researchers of the topic have made.

Preliminary research involves searching sources of theory and previous studies to discover what the appropriate bases for the subsequent, detailed work are likely to be – often, alternatives will be found. It is at this stage that the design of the main research must be formulated or confirmed.

In the dynamics of research, the process cycles through time, each new research project is able to build on those which have preceded it and *it is important* that they do so. Thus, it is essential that every researcher embarking on a project endeavours to discover what relevant work has been executed, as well as what theory bases apply, otherwise the wheel may be invented repeatedly and, without a base of theory, there will be little understanding of what has been done and the foundation from which progress may be achieved. Hence, progress in the development of knowledge is likely to be constrained.

Considerations of theoretical bases and previous research will shed light on appropriate methodologies to aid replication of studies and to approach the topic from alternative, but complementary, perspectives. Certain disciplines have traditions of employing particular methodologies and so, whilst a wealth of experience and expertise may have been accumulated, it may prove more difficult to establish the legitimacy of using a different methodology.

During the initial research phase, it will be useful to produce or, if produced already, to review the research model. Such a model will depict the main variables and the hypothesised relationships between them. Production of such a model begins at the conceptual level; that conceptual, or theoretical, model must be converted into an operational model – a model which can be used in practice to 'drive' the research and generate the variables which are to be observed and measured. In deriving the operational model from the conceptual one, inputs of previous research findings are employed – to determine what relationships (causalities) have been corroborated and which remain to be investigated.

Thus, the models identify what lies within the boundaries of the research project, known as the *endogenous variables*, and what lies outside the boundaries, called the *exogenous variables*. The 'permeability' of the boundaries determines the degree of influence *exogenous variables* may have on the system under investigation.

Data and information

Data, from an information system perspective, may be regarded as 'raw facts and figures' – measurements which can be made and recorded. Information is facts and figures which are expressed in forms suitable to assist a decision maker; information directly supports decisions.

> *Example*
> A contractor's tender sum for a building project is an item of data; the presentation of that sum in the context of other tenders for the project with a discussion of the 'levels' of the prices bid is information; hence, usually, information is raw facts and figures which have been 'processed'.

Data stand alone whilst information incorporates data and places them in context(s) – data are objective whilst information often contains subjective elements (judgements). People's answers to questions in which they are required to express opinions are forms of subjective data. Provided the responses have been obtained properly (see Chapter 6), the responses do constitute data.

Bechhofer (1974) considered the process of social research to be '... not a clear cut sequence of procedures following a neat pattern but a messy interaction between the conceptual and empirical world, deduction and induction occurring at the same time'. In examining the design of a research project, it is useful to consider the intended outputs from which the data collection requirements and the necessary analyses may be determined. Research may be regarded as an information system – the desired outputs, in terms of any hypothesis to be tested and the objectives to be realised, are the starting point for determining what is necessary in the other main parts of the system, the inputs and conversion process, given the operating environment.

It is important to consider all the main processes of the information system – the desired outputs, the available data and information, the required conversions; the feedback mechanism makes 'checking' viable as

well as allowing for development of the system; it should help to identify when and what environmental forces have an influence on the research
design.

'Mohr (1982) points out [that] there is a broad choice between variance and
process research designs. Variance designs are oriented towards the discovery and prediction of variance in phenomena of theoretical interest. Process
designs are oriented towards the discovery of the configurations and processes that underline patterns of association or change ... The tradition of
comparing national similarities and differences in organization adopts a variance approach. This now needs to progress beyond two serious limitations.
One is that the research conducted within this tradition has often examined
organizational characteristics only indirectly. Although more economical for
a given sample size, the use of either closed-ended scales, or of databases constructed for other purposes, is not a substitute for on-the-spot investigation
that is sensitive to both interpretative and objective definitions of the subject
matters ...'

'... Secondly, research designs will need to take a more comprehensive
view of context in order to locate units of study more precisely in relation
to the factors that potentially impact on their organization. Much previous
variance research failed to take account of the configuration of contextual
factors within each country, preferring instead to limit itself to selected economic, cultural or institutional factors. Without an adequate theory of how
these factors might themselves inter-relate, it has been tempting to ascribe
the organizational variance not predicted by the selected variables to "noise"
ascribed in a non-theorized manner to other ill-defined variables. These were
treated merely as theoretically mysterious residuals. Thus what economic and
technological contingencies failed to predict was often ascribed to "culture"
without any theoretical justification.' (Child 1981)

To continue, '... future variance research will need to employ quite elab-
orate research designs that fulfil several conditions. There should be a more
comprehensive theorization of both independent and dependent variables
than has hitherto been typical. The theorizing should refer to both low and
high context perspectives and be articulated in advance through hypotheses or other means. Guided by such theorization, cases will have to be
selected with careful attention to how they are situated vis-à-vis the local
and global factors hypothesized to have a potential impact on their organization. Account has to be taken of within-nation as well as of between-nation
variance in contextual features.' This specificity in respect of context is also
commended by Earley and Singh (1995) in terms of what they call 'the hybrid
approach, which combines a comprehensive overview of the systems in which
firms operate with examination of the specific inner workings of the systems
themselves.'

Dynamic process

'A process approach to research is oriented towards change and development' (Bhagat & Kedia 1996). Child (1999) comments on the same subject:

'It is concerned with the potential dynamics over time between material and ideational forces and low and high context factors, and how these relate to organizational structure and processes. It would therefore call for longitudinal research designs or at least as that permitted insight into the impact of different forces on ongoing developments such as the process and rationales of decision making about organizations.'

'. . . moves towards theoretical integration are handicapped by a lack of conceptual consistency. This takes two forms. The same concept such as control, is defined in a variety of different ways. The second methodological challenge encountered in studying organizations cross-nationally is therefore to find ways of further underpinning the integration of different theories by increasing the operational equivalence of their commonly employed concepts.'

'The intention is to arrive at a multidimensional operationalization of the concepts that not only takes into account those aspects supposed to be prominent within a given culture but also permits an exploration of the possible overlap and similarities between dimensions emphasised by different cultures.'

Pettigrew (1997) suggests that the '. . . purpose of process analysis . .[is]. . to account for and explain the what, why and how of the links between context, process and outcomes . .[there are]. . five internally consistent guiding assumptions:

(1) embededness, studying processes across a number of levels of analysis;
(2) temporal interconnectedness, studying processes in the past, present and future time;
(3) a role in explanation for context and action;
(4) a search for holistic rather than linear explanations of process; and
(5) a need to link process analysis to the location and explanation of outcomes'.

> *Example*
> Bresnen *et al.* (2005) use processual analysis comprising a number of longitudinal case studies of construction organisations to research the impacts of organisational structures and agency roles on implementations of changes.

Initial search

An essential early stage of virtually all research is to search for and to examine potentially relevant theory and literature. Theory and literature are the results of previous research projects. Theory is the established principles and laws which have been found to hold, such as Einstein's theory of relativity; theories of the firm. Thus, 'A theory is a statement of relations among concepts within a set of boundary assumptions and constraints' (Bacharach 1989). Literature, in this context, concerns findings from research which have not attained the status of theory (principles and laws); often, it represents findings from research into particular applications of theory.

The items of theory and major references should be established in early discussion with the supervisor and others who are experts in the topic. Consultations to determine the usefulness of the proposal during its formulation will reveal appropriate theories and major research projects which have been carried out. These are good starting places, but it should be remembered that references are always historic although research journals, including leaflets published by research councils, private research organisations and professional institutions, often publicise current research projects. Employment advertisements for research posts often note the projects to be undertaken and so, indicate where interest and expertise in those topics can be found. The world-wide web is a very popular, efficient and easy-to-use tool for finding experts and information. However, there is no guarantee of authenticity, reliability and validity and so, indicators must be employed (employing organisations for persons; authors and 'hard copy' publication media etc.) for data and information.

Fortunately, despite work pressures, most researchers are keen to collaborate and help others' investigations, but it is important that what is asked of such experts is well focused and shows reasonable knowledge of the subject and the desire to investigate a topic of some recognisable importance. A blanket request, all-too-common via email, to 'tell me everything you know about . . . quickly and for free', will not be welcomed.

Research papers, which constitute the largest and the most important wealth of literature available, usually include a review of theory and literature which informed and underpinned the work reported, including the methods used. So, in proceeding to note data collected, the analyses executed, results obtained and conclusions drawn, research papers present distillations of previous work on the topic and advances made by that piece of research itself.

In considering the research process, attention should be given to 'DATA'; namely:

D **D**efinitions of the main terms involved; especially where terms have varied definitions, it is essential to decide *explicitly* the definitions to be adopted, and why they have been adopted, so that critical literature review and appropriate data collection can be made.

A note the **A**ssumptions which are made and the justification for them; if possible, explain their consequences and examine what occurs when any assumptions are relaxed.

T research, and critically review the **T**heories, principles and literature relating to the subject matter of the research.

A evaluate what **A**nalyses may be carried out with respect to data available, the objectives and any hypothesis, so that the most robust and rigorous analytic methods will be used, thereby maximising confidence (validity and reliability) in the results.

Definitions and assumptions

Reflect upon the topic. The draft proposal and model of variables almost certainly will contain terms to which particular meanings must be attributed.

> *Example*
> Often, industry contexts and meanings are *assumed* to apply and to be known and accepted widely – what is a 'frog'? In a bricklaying context it is a hollow (void) on one face of a brick; to the general public, it is a small, greenish amphibian which hops and croaks.

Therefore, it is a good principle to identify assumptions and to define terms – management contracting and construction management are different in North America from the procurement routes given those names in UK. Literature is valuable in establishing the variety of terms and definitions, which are important points of debate, and hence one can select the definition most appropriate for use in the research project. Usually, a good starting place for deciding definitions of terms is the *Oxford English Dictionary*, which provides the basic, normal definition(s), further refinement may be required to define terms in particular subject or practice contexts – for such refinements, or particularisations, it is important that authoritative sources are used.

Before progressing, it is good practice to review the proposal with the supervisor of the research and with colleagues. The review should attempt to ensure that the assumptions and the important terms have been noted clearly and defined appropriately for the intended work. The test is whether the proposal can be understood unambiguously by any intelligent person; not just someone very familiar with the context and the particular topic of the proposed research.

Theory and literature review

The definition of the topic and terms must have been established during the production of the research proposal; the programme of work will show the time available, although it is usual for the review to be 'kept open' so that further literature can be incorporated if any work of significance is discovered during later stages of the project; however, care is required. Although keeping the review open in that way is useful to ensure it is comprehensive and to incorporate latest research findings, it may lead to the review never being finished – so it is important to establish a 'final deadline' to close entries to the review. Literature should not merely be found and reviewed, the body of relevant literature from previous research must be reviewed *critically*. The literature must be considered in the context of theory and other literature – the methodologies, data, analytic techniques, sampling etc. – so that objective evaluation takes place.

Neither in considering theory nor in the critical review of literature is it appropriate for the researcher to express personal opinion – let the theory and/or the literature 'do the work'. Alternative views and findings must be abstracted and ordered, to present thematic discussions such that a coherent debate is presented through synthesis and evaluation. Bodies of acceptance should be categorised as should arrays of issues – both of which inform subsequent portions of the research. Often, weighing of arguments in the context of theories and, if applicable, the theoretical stance adopted for the work is required. This provides and demonstrates appreciation and understanding of the state of knowledge of the topic and its context.

According to Haywood and Wragg (1982), the literature review must be critical and, therefore, demonstrate that '. . . the writer has studied existing work in the field with insight'. The insights should be derived from both the theoretical considerations and the completeness of the review of the literature. A mere listing of the articles which have been read with a summary of their main points is *not sufficient*; the critique – drawing out issues and arguments, setting alternative views against each other etc. – is the essential element of the *critical review*.

Therefore, the review of theory and literature must provide the reader of the research report with a summary of the 'state of the art' – the extent of knowledge and the main issues regarding the topic which inform and provide rationale for the research which is being undertaken. Naturally, it is useful if the review not only informs the reader of the basics of the research, but makes the reader eager to read the subsequent sections of the report.

Analysing data from a search

A random search is unlikely to reveal much of significance for the topic – it is important that the search is structured. Often, it is best to begin with the theory on which the research is based – leading texts will be a good place to start as they provide statements and explanations of the theory and references to other work citations. Often, the references are useful in providing an introduction to the literature – seminal papers which, themselves, provide lists of further references. Citation publications (indexes) are helpful in tracing links between publications, especially journal papers.

As the scope of titles of publications can be enormous, even on a single, well-defined topic, it is helpful to have several approaches to discovering information, i.e. adopt a triangulated search. It is useful to list theories to be considered, leading authors and topic keywords. The more precise the search keys can be, the more both time and expense can be saved. Many online searches are expensive and so, commonly, libraries restrict the number of words which can be entered for a search. It is a good idea to do some preliminary investigations to ensure that the best use of such facilities is being made.

Naturally, the lists of authors will grow and grow; that should not happen to the list of keywords – the list of keywords may change but it is a good idea to limit its size to a small number; many research journals (etc.) limit the number of keywords authors can use to a maximum of five or six, and a maximum of three is not unknown. So, *good topic definition is essential.*

> *Example*
> Consider writing a list of the libraries which will be useful to consult or visit to obtain information for research.
>
> Apart from containing the local university and college libraries, the list should contain other local libraries and specialist libraries – of professional institutions (CIOB, RICS, ICE etc.) and research organisations. Some large companies and consultants' practices have libraries of their own. The British Council local offices often have a limited library facility. More importantly, a link to the British Library (with its vast collection) can be available through the interlibrary loan facility.

Obtaining access to libraries may present some difficulties, so it is best to enquire by telephone or letter before making a visit, especially if the library is some distance away. Most public, university and college libraries allow access for reading but may restrict borrowing. However, a charge, even for reading facilities, is made in some countries and it is useful to carry evidence of status as a research student of a university etc. to gain both access and assistance. Often, private libraries have more restricted access, such as to people with membership of the professional institution involved, although, by prior arrangement and evidence of the research status of the applicant, access can be obtained in most cases, it just may take a little time!

Usually, library staff are very helpful. However, an efficient search pattern is likely to involve consulting the local university subject librarian for initial advice and assistance and, once the primary information has been obtained and scrutinised, referring to specialist libraries for particular items and topics to complete the picture.

Increasingly, libraries employ electronic technology – computer databases have replaced card indexes, which makes searches much quicker, easier and more comprehensive. It is helpful to consult abstracting services as a mechanism for preliminary selection of what papers to obtain – a title may not indicate the paper's contents accurately but an abstract, even if very brief, should provide a valuable synopsis of content (subject matter and method(s) of investigation) and conclusions. Especially where particular researchers are 'key' to progressing in a topic (such as Milton Friedman and Monetarism), citation indexes are useful sources to trace developments in the topic via papers and researchers who have cited (made reference to) the 'key' individual. Clearly, the detective work in collecting literature involves a considerable amount of logical networking. *Do not lose sight of the boundaries of the topic.* Ensure what is collected is relevant to the aim, objectives and (if appropriate) hypothesis.

The increasingly widespread availability of the internet and its well-known search engines is both a boon and a burden. As persons become more and more accustomed to using the internet to search for and obtain information, without having to move from their desk (or armchair/bed), with no more physical effort than depressing computer keys, and with no more mental effort than accessing a search engine and determining reasonably appropriate keywords for the search, this amazing facility seems to be one of the greatest aids for research yet! However, it creates problems too concerning both the quantity and quality of information available and obtained. Research requires relevant, valid and reliable information.

The quantity of information yielded by many internet searches is huge and not of guaranteed quality. Hence, the researcher must assess each item to evaluate its relevance, validity and reliability before using it.

If not developed to some extent by this stage of the work, speed reading and writing are useful skills. Researchers need to reach a large quantity of material and abstract all potentially helpful items *with accurate references*. It may be helpful to provide the basic reference to the paper etc. at the top of a sheet and to note page numbers in the margin as brief quotations and summaries of contents are abstracted. As it may not be possible to borrow all the books or journals required, photocopying relevant sections may be possible – be careful in doing so that copyright is not infringed. Some sources may not be available immediately – books may be on loan, probably with a waiting list, copies of papers from other libraries etc. will take time to arrive. Whilst such time requirements may not be a major problem, they must be recognised and incorporated into the work programme. So, the earlier material is requested the better, but this possibility is dependent on how well the research topic has been defined.

When nearing completion of the collection of theory and literature or as the deadline for completion of that part of the research approaches, as indicated on the programme, the information should be categorised and ordered in a logical sequence to present the basis of the research in the report. Reference to the proposal which was produced will be helpful and, although alternative presentations are available, certain principles apply irrespective of which alternative is adopted.

As research builds on theory, it is helpful to provide a review of the theory before presenting the relevant literature of research findings. Theory and literature may be presented in a series of topic or sub-topic categories or the array of theory followed by the literature. To an extent, the issue is one of personal choice, but for larger projects involving a diversity of theory, it is useful to present theory and literature category by category with an overall summary review; this approach aids both flow of argument and understanding by a reader.

Example

Consider a research proposal on the topic of bidding for construction work (whether by contractors to win projects or by clients to let projects, including maintenance work etc.). Preliminary lists of aspects of theory and literature, including leading texts and research papers to be explored, might include:

Theory: Competition
Corporate objectives and strategies
Behaviour – individuals and organisations

Literature: Friedman (1956)
Gates (1967)
Carr (1977)
Hillebrandt & Cannon (1990)
Langford & Male (1991)
Drew & Skitmore (1997)
Rooke *et al.* (2003, 2004)

It is useful to keep the lists available for updating and extending as the initial research progresses so that, on completion, the lists of theory and literature in support of the research will be comprehensive.

Assembling the theoretical framework

Theory provides the framework for the research project rather like a structural steel or reinforced concrete frame is used in a building. It will also indicate the data which should be collected and further theory will denote appropriate methods and techniques of analysis.

It is essential that theories themselves be subject to rigour of analysis. Especially in social sciences, it is quite common to encounter theories which are in conflict. Principles and laws in these disciplines are derived from observations and analyses of human behaviour, so the complexity of integrating findings etc. leads to varying interpretations of behaviours and substantiations of those interpretations by further testing. The dynamism of societies complicates the research extensively – people's behaviours vary for many reasons; gravity and other physical and natural laws are reasonably invariate, whilst our understanding of such laws is developing still.

So, bodies of theory must be examined and evaluated to arrive at a theoretical basis or framework appropriate to the research proposed (the research paradigm). It may not be possible to decide the logical body of theory to use from the description of theory provided, and it will not be possible to weigh alternative and possibly competing theories. In such a situation a choice must be made. The basis of such a choice may be personal preference caused by familiarity or expertise with an approach or set of ideas, a sympathy with the theoretical perspective (e.g. a Keynesian, a Monetarist or a Marxist view of inflation), or findings from leading research in the topic.

Of course, it is debatable whether competing theories can constitute basic 'principles and laws', or whether they are perspectives and beliefs which give rise to partly supported hypotheses. Perhaps from a pragmatic perspective of a researcher using the theories and irrespective of what they are from a research philosopher's point of view, the 'true' nature of the theories is relatively incidental *provided* their natures are recognised and taken into account in the execution of the research project. The question is one of academic contention – especially in terms of the philosophy of research and the validity of findings; what are realities, what are truths and how do they differ? At this stage, it is important and sufficient to be aware of issues and to recognise and use theories for what they are – limitations and all.

Therefore, the theory adopted provides the basic structural framework to identify and explain facts and the relationships between them. Consider a diagram of the variables in a research proposal: determination of the variables, identities of the variables, and the relationships between them should be determined from theory. Hypotheses are employed to fill in gaps – to suggest relationships which may exist if theory is 'extended', or to relate two aspects of theories. Where theories are really partly supported hypotheses, those hypotheses are appropriate for further testing so, for example, it is likely to be appropriate to investigate which theory of inflation is more appropriate in a particular situation.

It is a good idea, whenever possible, to use theory to build a model of the proposed research – the variables and relationships, the points of issue and those of substantiation. In cases where assumptions are involved, maybe as a requirement or pre-condition for the theory or law to hold, such as Boyle's Law, it is essential to note the assumptions *explicitly* as they impose limitations on the research and its findings as well as prompting questioning and, perhaps, investigation of what occurs when the assumptions are relaxed. The main things to be done and their sequence, to assemble the review of theory and literature into a theoretical framework, include:

- Defining the topic and terms; time and cost limitations.
- Noting items of theory.
- Noting major references known.
- Listing keywords for theory and the topic.
- Locating libraries.
- Obtaining access to libraries.
- Investigating library resources – abstract and citation indexes, computer search facilities etc.
- Executing searches.
- Obtaining and reviewing sources with concise recording of information.
- Assembling the review.

Philosophy and methodology

Many people are prone to use the term methodology as equivalent to the scientific empirical approach which addresses (1) the definition of the problem – defining the problem in a such a way that it lends itself to careful investigation; (2) the statement of the problem and its interpretation – the statement can be interpreted in the light of theory and accumulated knowledge; (3) the formulation of hypothesis – the hypothesis must stem from the body of theoretical principles and expressed unambiguously so that it can be tested; (4) the empirical testing of the hypotheses – the findings of which would modify the existing body of scientific knowledge.

Table 3.1   Polarities and assumptions adopted by researchers.

	Continuum	Assumptions
Ontology	Nominalism–realism	Whether the object of investigation is the product of consciousness (nominalism) or whether it exists independently (realism).
Epistemology	Anti-positivism–positivism	What our grounds of knowledge are.
Human nature	Voluntarism–determinism	Whether humans interact creatively with the environment (voluntarism) or whether they are passive objects (determinism).
Methodology	Ideographic–nomothetic	Ideographic ('concrete') or nomothetic (abstract) approaches to evidence collection.

Adapted from Remenyi *et al.* 1998, p. 103.

In considering the appropriate research design, researchers must consider (and make known) to which research community they believe they belong as well as the epistemological, ethical and ontological assumptions of their research (Remenyi *et al.* 1998). The deductive approach to research has become synonymous with positivism (Gill & Johnson 1991). Burrell and Morgan (1979) discuss the assumptions which researchers make regarding the polarities of the subjective–objective continuum (mirrors the positivist–non-positivist continuum) and researchers may consider adopting any strategies along the continuum (Table 3.1).

In considering the philosophy and methodology of research, which is a complex topic, one considers the principles that guide the process in extending knowledge and seeking solutions towards the research problems. For instance, the researcher considers the issues of ontology and epistemology, constructivism, reductionism, positivism, phenomenology etc., all of which affect the statement of the research problem and the subsequent research design.

Ontology and epistemology

Ontology concerns with the assumptions in conceptual reality and the question of *existence* apart from specific objects and events, e.g. the ontological argument regarding the existence of God. Epistemology is the branch of philosophy that concerns the origins, nature, methods and limits of human knowledge.

Positivism and phenomenology

While positivism and phenomenology represent different philosophical bases, they are not mutually exclusive in the adoption of methodology. While there has been a paradigm shift towards the more qualitative, interpretivist approach in construction management research, the inherent difficulties, and potential shortcomings and biases, in this approach must be acknowledged. For instance, ethnocentrism, in interpretation of cultural norms and observable behaviours, is, particularly, an area of concern, especially in cross-cultural studies. Even in apparently homogenous cultural groups (group boundary is a matter of definition), the knowledge/bias/assumptions of the interpretivist/researcher is likely to affect the conclusions drawn and their validities.

'Interpretation' of observed phenomena is greatly emphasised in the interpretivists' approach. The very act of interpretation 'implies the existence of a conceptual schema or model on the part of the interpreter such that what is being observed and interpreted is assumed to conform logically to the facts and explanations inherent in the model' (Reber 1995). There can be two types of interpretations, cognitive interpretation and scientific interpretation. The model in cognitive interpretation is the mental schema within which all stimuli are identified and reacted to. The model in scientific interpretation is a theoretical one which seeks to explain/characterise reality. When we interpret a response or the behaviour of a group of people, the process involves induction and generalisation from some scientific schema/model.

Weber (1978, vol. 1, p. 9) is emphatic that persuasive interpretation of social action is necessary but not sufficient since 'every interpretation attempts to attain clarity and certainty, but no matter how clear an interpretation as such appears to be from the point of view of meaning, it cannot on this account claim to be the causally valid interpretation'. Both interpretation and scientific verification are essential and possibilities should be sought in using both quantitative and qualitative methods to pursue explanations adequate at the levels of both cause and meaning.

> *Example*
> An example given by Remenyi *et al.* (1998, p. 59) is that in-depth interviews, which can provide good results if used in association with an interpretivist view of evidence analysis, can also be used with content analysis, which is, essentially, a positivistic way of assessing evidence. Hence, it is possible to adopt a research design that is, essentially, positivist in nature in a phenomenological manner. Similarly, a phenomenological approach can be used to produce a positivistic view of the subject being studied.

Physical and natural scientists mostly take a positivistic approach. The philosophical stance sees the positivist researcher as an objective analyst and interpreter of a tangible social reality, i.e. assuming that the researcher is independent of and neither affects nor is affected by the subject of the research. Causality is assumed and quantifiable observations/evidence is critical, parsimony is important and it should be possible to generalise or to model the observed phenomena. According to Popper (1989), the central tenet of positivism is the idea of falsification, i.e. an idea cannot be regarded as scientific unless it is falsifiable, though such scientific verification is thought to be naïve by Lakatos (1970).

Phenomenology advocates the scientific study of immediate experiences and focuses on events, occurrences and happenings as one experiences them, with a minimum of regard for the external, physical reality. Phenomenological analysis avoids focusing upon the physical events themselves but deals with how they are perceived and experienced. Interpretation of meaning is derived by examining the individual's relationship with and reactions to these real-world events.

Sociologists would argue that phenomenology (non-positivistic approach) is appropriate for studying people and their organisations. Cohen and Manion (1987) define phenomenology as a theoretical point of view that advocates the study of direct experience taken at face value; and one which sees behaviour as determined by the phenomena of experience rather than by external, objective and physically described reality. Perhaps the essence of phenomenology is that it is 'a term that carries a great deal of ambiguity along with its sometimes confused and faddish use' (Boland 1985). To understand phenomenology, the primacy of context must be acknowledged. 'People cannot be understood outside of the context of their ongoing relationships with other people or separate from their interconnectedness with the world' (Clarkson 1989, p. 14). At the heart is the relationship between self and society (Mead 1934) as the 'organism creates its environment' (Clegg & Dunkerley 1980, p. 267).

According to Remenyi et al. (1998, p. 87), phenomenology is a philosophical stance and not all qualitative researchers necessarily subscribe to it. 'The term phenomenology essentially describes the philosophical approach that what is directly perceived and felt is considered more reliable than explanations or interpretations in communication. It is a search for understanding based on what is apparent in the individual environment rather than on interpretations made by the observer.'

> *Example*
> For instance, Marshall and Rossman (1995) distinguish between phenomeno-
> logical interviewing and other non-positivist interviews used to gather qualitative
> evidence; Remenyi *et al.* (1998) distinguish a feminist researcher, who places
> less stress on the primacy of the respondent experience, and the phenomenol-
> ogy researcher who makes specific attempts to remove all traces of personal
> involvement in the evidence collection exercise. A feminist interviewer's interest
> is in the 'validation of women's subjective experiences as women and as people'

Example   Continued
(Oakley 1981, p. 30) and rejects the formalism imposed on research activity by a positivist approach. This calls into question the extent to which the research process can be governed by objectivity on the part of the researcher. The same concern is applicable in culture research.

Researchers' interpretations of observations are subject to error and bias, hence, the issue of validity is important. Bias can be cautioned against but, in many cases, cannot be removed. The relativist position is that there are only truths but no universal truth, versions of reality but no one reality. Our views are affected by culture and, 'Different cultures employ radically different conceptual schemes defining what exists in the world, how things are organised in time and space, what sorts of relation obtain among things, and how some things influence others . . .' (Little 1991, p. 203).

Example
Often, it is said that individuals from a collectivist society have greater emotional dependence on one another than individuals from individualist societies. Earley (1997, p. 144) gives an example that 'in Asian cultures, an individual's company is expected to provide not only a salary, medical coverage, and other benefits common to the West but also housing, child care, education, and even moral and personal counselling as well as political indoctrination'. However, in Hong Kong (certainly prior to 1997), Earley's allegations of the above benefits in Asian cultures were, generally, not present amongst the local Chinese but were inherent in the social system of the expatriate culture. So, what is Asian culture (when interpreted through a Western researcher's lenses of observable phenomena) from the perspectives of the positivist and phenomenologist?

Constructivism and reductionism

The essence of all constructivist theories is that perceptual experience is viewed as more than a direct response to stimulation, i.e. it is an elaboration or construction based on hypothesised cognitive (reasoning) and affective (emotional) operations.

Example
In social psychology, social constructivism approaches the study of social psychological topics from the same philosophical stance. Social constructivists argue for

continued

> *Example Continued*
> the notion that there is no such thing as a knowable objective reality; rather, they
> maintain, all knowledge is derived from the mental constructions of the mem-
> bers of a social system (Reber 1995). Construction management researchers
> often view the project organisation as a social–technical system and, in the study
> of the relationships among project participants, take the social constructivist's
> approach.

Broadly speaking, reductionism is a philosophical point of view which maintains that complex phenomena are best understood by a componential analysis
which breaks down the phenomena into their fundamental, elementary
aspects. The core of the reductionist's position is that greater insight into nature
will be derived by recasting the analyses carried out at one level into a deeper,
more basic level (Reber 1995).

> *Example*
> Contemporary behaviourists seek to reduce all complex acts to stimulus–
> organism–response (S–O–R) terms. Cognitive theorists maintain that images,
> thoughts, plans and ideas exist as entities with causal roles to play in behaviour
> whose reality is distorted by a re-casting into an S–O–R format.

Fuzzy thinking

Binary precision is part of the scientific method and logical positivism remains
the dominant philosophy of modern science and engineering. Aristotle's logic
lies behind the bivalent approach in reasoning, e.g. *A or not-A*. However, the
world is not entirely black and white and the logical contradiction in bivalence happens in *A and not-A*. Fuzzy logic confronts this 'either-or' concept;
fuzziness begins where contradictions begin, where *A and not-A* holds to any
degree.

The fuzzy philosophy states that everything is a matter of degree – a world
of multivalence and the opposite of which is bivalence (the black and white).
Positivism demands evidence, factual or mathematical. Based on binary logic,
it comes down to one law: *A or not-A* – it cannot be both *A and not-A*; for
instance, the sky cannot be both 'blue' and 'not blue' at the same time, thus
ignoring shades of blue. Fuzzy logic is reasoning with fuzzy sets. A fuzzy
cognitive map is a fuzzy causal picture of the world and a fuzzy system is a
set of fuzzy rules that converts inputs into outputs.

Eastern thoughts (e.g. Tao) offer the belief systems that accept contradictions
and fuzziness, systems that work with *A and not-A*, with yin and yang. (The

Tao symbol actually shows the black within the white and the white within the black – in constant motion.) In the fuzziest case, the glass of water is as half empty as half full, where yin equals yang (in the ancient Tao philosophy). According to Kosko (1994), 'the yin-yang symbol is the emblem of fuzziness' which depicts a world of opposites and 'fuzzy logic begins where western logic ends' (Kosko 1994, pp. 14–17). The fuzzy approach offers an alternative to positivism (Kosko 1994; Zadeh 1965; Zimmermann 2001).

> *Example*
> Elbeltagi *et al.* (2001) use fuzzy logic to investigate how planning the layouts of construction sites depends upon the scheduling of the execution of the construction operations.

Theoretical models and constructs

What is modelling?

Modelling is the process of constructing a model, a representation of a designed or actual object, process or system, a representation of a reality. A model must capture and represent the reality being modelled as closely as is practical, it must include the essential features of the reality, in respect of the purpose of constructing the model, whilst being reasonably cheap to construct and operate and easy to use.

> *Example*
> Models occur in a variety of forms and serve many purposes. A toy car, which must be a tiny replication (in appearance and certain other basic features – rotating wheels etc.) of the real car which it represents, enables a child to learn through play – often play which simulates the actions of a real car in situations both experienced and imagined by the child. Toy building bricks and architectural models offer different levels of sophistication (detail, complexity, accuracy) in representing buildings, whilst mathematical models are employed by engineers in the design of structural components and building services systems. Economic and econometric models are used extensively – project cash flow models, models of resource inputs required for different types of projects, the Treasury's models of the UK economy. Population models are used to forecast demand in different sectors of the economy. Another main function of models is to facilitate reasonably accurate prediction, such as programming techniques to produce models of entire project durations and component activity sequences and durations.

The objective of the model should be to reflect the purpose of the model, such as the questions to which answers are to be sought from using the model. One should know for whom the model is to be constructed, in order to lend perspective to the modelling and to suggest sources of data, forms of outputs etc. The analysis stage comprises organised, analytic procedures to determine the operation of the reality, noting the location and permeability of the boundary of the system to be modelled. Often, a diagram of the reality will be of benefit in identifying variables and their relationships prior to the quantification of both. This is a major element of the synthesis stage, which yields one model or an array of alternative models. It is likely that the resulting models will reflect both the education and training of the analysts and their experiences of modelling other, especially similar, realities.

Theoretical model

Often, the researcher establishes a research model from the review of literature and theories (theoretical model) which then forms the basis for setting hypothesis and testing the relationships of variables. A theoretical model is a set of variables and their interrelationships designed to represent, in whole or in part, some real system or process. Common forms of theoretical modelling in construction research are graphical models and mathematical models. Graphical models are visual and logical; they suggest directions of relationships among the variables but they do not provide numerical results. The graphical models can provide preliminary steps to developing mathematical models. Mathematical models explicitly specify the relationships among variables in mathematical form.

The formulation and construction of a theoretical model requires a variety of inputs, as does any research activity. Following the determination of the objectives and limitations of the research, the initial stage is to investigate existing theory and principles. Once appropriate theories and principles have been distilled from the existing body of knowledge, literature can be searched to identify the applications of these theories and principles and findings thereof, in research projects and in practice. Such investigation will indicate appropriate variables to define, isolate and measure (usually via experimentation, whether in a laboratory environment or in-use in a 'reality'), so that performances of the individual variables and their relationships can be evaluated.

> *Example*
> The example of Time Series Analysis and Forecasting, in Chapter 7, may be regarded as an instance of elementary modelling.

Constructs

The variables require *theoretical* and *operational constructs* to be established in order to build the model. Theoretical construct concerns conceptual definition – which is an abstract definition derived from theories and is contrasted with an operational definition. Operational definition (is used in defining an operational construct) specifies precisely how a variable is measured in a particular study.

Much construction management research involves the measurement of hypothetical constructs (or theoretical constructs), e.g. motivation, learning etc. It can be said that these entities are hypothesised to exist on the basis of indirect evidence (Leary 2004, p. 70) and researchers often examine construct validity by calculating correlations between the measure of the construct and scores on other measures (Cronbach & Meehl 1955).

Once the structure of the model has been established and its performance scrutinised and determined to be suitable for the objectives, appropriate values can be input for the necessary variables and the resultant outputs calculated. Clearly, the direction of using a model may not be the same as was employed to construct, verify and validate the model.

> *Example*
> In structural engineering, the performance of a member may be modelled by constructing various test members (for experimentation) of different sizes and combinations of components of known properties (tensile strength of steel, compressive strength of concrete etc.) to establish its load-bearing capabilities, performance characteristics and failure modes. The resultant model then may be used (with appropriate 'factors of safety') to design the components required for the member to achieve the necessary performance characteristics.

There is a strong relationship between methodology and the theoretical model. For instance, a theoretical model can be based on soft systems methodology which adopts the philosophy of social constructivism. Green and Simister (1999) consider a social constructivist approach to systems modelling, following Checkland's (1981, 1989) and Checkland and Scholes' (1990) development of soft systems methodology (SSM). Essentially, the progression is from hard systems, which regard the system as existing in the real world, often as a static, technical mechanism, to soft systems, which incorporate the social dimensions in a dynamic world. Fig. 3.3 shows the process of modelling by relating the 'traditional' modelling process (e.g. Mihram 1972) and sampling to the approach of SSM.

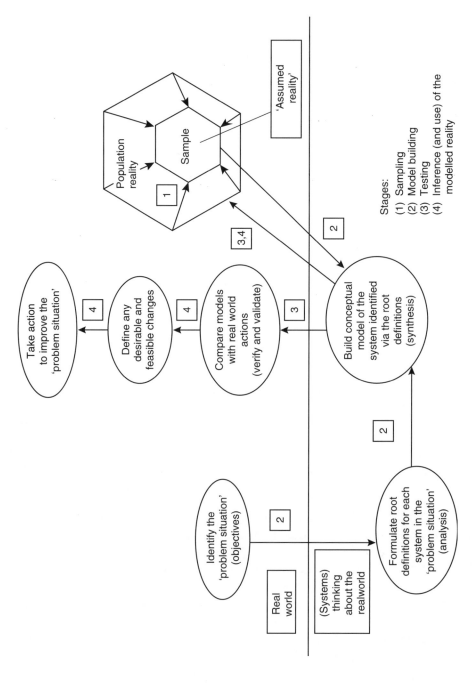

Fig. 3.3   The modelling process (developed from Taha 1971; Checkland 1989).

Proper referencing

If sections of the work are not referenced, they are assumed to be original work of the researcher. If that is not the case and the work, ideas, etc. have been obtained from someone else, presenting that work as the researcher's own is *plagiarism* – intellectual theft – and it is treated very seriously indeed. However, virtually everyone makes mistakes and occasionally omits references accidentally. Plagiarism is omitting references deliberately; it has been found to the extent of copying and submitting someone else's entire dissertation. In such instances condemnation will be swift and complete.

There are several standard methods for referencing. The Harvard system is used very widely and is gaining popularity. Most publications and institutions prescribe the referencing system to be used.

Example
The Harvard system of referencing is:

Textbook:
In the text of a report:

Quotation: 'What was the relationship between theory, data collection and analysis?' (Burgess 1984)

Paraphrase: Burgess (1984) posed the question of the relationship between theory, data collection and analysis of those data.

In the references section of the report:

Burgess, R.G. (ed.) (1984) *The Research Process in Educational Settings: Ten Case Studies*, Falmer Press, Lewes

Journal:
The entries in the text of the report are as shown for textbook.

In the references section of the report:

Roth, J. (1974) 'Turning adversity into account', *Urban Life and Culture*, **3**(3), 347–359.

Summary

In this chapter, we have considered the requirements during the early stages of carrying out the research work – the collection and reviewing of theory and literature. Mechanisms to assist collection of such information have been discussed; in particular, the necessity is to be systematic and rigorous. It is essential that the review produced in the research report is a critical review and is referenced correctly.

References

Bacharach, S.B. (1989) Organizational theories: some criteria for evaluation, *Academy of Management Review*, **14**(4), 496–515.

Bechhofer, F. (1974) Current approaches to empirical research: some central ideas. In: *Approaches to Sociology: An Introduction to Major Trends in British Sociology* (ed. J. Rex), Routledge & Kegan Paul, London.

Bhagat, R.S. & Kedia, B.L. (1996) Reassessing the elephant: directions for future research. In: *Handbook for International Management Research* (eds B.J. Punnett & O. Sheikar), Blackwell, Cambridge, MA.

Boland, R. (1985) Phenomenology: a preferred approach to research on information system. In: *Research Methods in Information Systems* (eds E. Mumford, R. Hirschheim, G. Fitzgerald & T. Wood-Harper), Elsevier Science Publishing Co., Amsterdam.

Bresnen, M., Goussevskaia, A. & Swan, J. (2005) Implementing change in construction project organisations: exploring the interplay between structure and agency, *Building Research and Information*, **33**(6), 547–560.

Bryman, A. & Cramer, D. (1994) *Quantitative Data Analysis for Social Scientists* (revised edn.), Routledge, London.

Burrell, G. & Morgan, G. (1979) *Sociological Paradigms and Organisational Analysis – Elements of the Sociology of Corporate Life*, Heinemann, London.

Carr, R.I. (1977) Paying the price for construction risk, *Journal of the Construction Division, Proceedings of the American Society of Civil Engineers*, CO1, **103**(March), 153–161.

Checkland, P.B. (1981) *Systems Thinking, Systems Practice*, Wiley, Chichester.

Checkland, P.B. (1989) Soft systems methodology. In: *Rational Analysis for a Problematic World: Problem Structuring Techniques for Complexity, Uncertainty and Conflict* (ed. J. Rosenhead), pp. 71–100, Wiley, Chichester.

Checkland, P.B. & Scholes, J. (1990) *Soft Systems Methodology in Action*, Wiley, Chichester.

Child, J. (1981) Culture, contingency and capitalism in the cross-national study of organizations, *Research in Organizational Behaviour*, **3**, 303–356.

Child, J. (1999) *Theorizing about Organization Cross-nationally*, seminar essay, University of Hong Kong, November.

Clarkson, P. (1989) *Gestalt Counselling in Action*, Sage Publications, London.

Clegg, S. & Dunkerley, D. (1980) *Organisation, Class and Control*, Routledge and Kegan Paul, London.

Cohen, L. & Manion, L. (1987) *Research Methods in Education*, Croom Helm, London.

Cronbach, L.J. & Meehl, P.E. (1955) Construct validity in psychological tests, *Psychological Bulletin*, **52**, 281–302.

Drew, D. & Skitmore, R.M. (1997) The effect of contract type and size on competitiveness in bidding, *Construction Management and Economics*, **15**(5), 469–489.

Earley, P.C. (1997) *Face, Harmony, and Social Structure*, Oxford University Press, New York.

Earley, P.C. & Singh, H. (1995) International and intercultural management research; what's next?, *Academy of Management Journal*, **38**, 327–340.

Elbeltagi, E., Hegzay, T., Hosny, A.H. & Eldosouky, A. (2001) Schedule-dependent evolution of site layout planning, *Construction Management and Economics*, **19**(7), 689–697.

Friedman, L. (1956) A competitive bidding strategy, *Operations Research*, **4**, 104–112.

Gates, M. (1967) Bidding strategies and probabilities, *Journal of the Construction Division, Proceedings of the American Society of Civil Engineers*, CO1, **93**(March), 75–107.

Gill, J. & Johnson, P. (1991) *Research Methods for Managers*, Paul Chapman Publishing Ltd., London.

Green, S.D. & Simister, S.J. (1999) Modelling client business processes as an aid to strategic briefing, *Construction Management and Economics*, **17**(1), 63–76.

Haywood, P. & Wragg, E.C. (1982) *Evaluating the Literature*, Rediguide 2, School of Education, University of Nottingham, Nottingham.

Hillebrandt, P.M. & Cannon, J. (1990) *The Modern Construction Firm*, Macmillan, Basingstoke.

Kosko, B. (1994) *Fuzzy Thinking. The New Science of Fuzzy Logic*, Harper Collins, London.

Lakatos, I. (1970) The methodology of scientific research programmes, *Philosophical Papers Volume 1*, Cambridge University Press, Cambridge.

Langford, D.A. & Male, S.P. (1991) *Strategic Management in Construction*, Gower, Aldershot.

Leary, M.R. (2004) *Introduction to Behavioral Research Methods*, 4th edn, Pearson Education, Allyn and Bacon, MA.

Little, D. (1991) *Varieties of Social Explanation*, Westview, Boulder, CO.

Marshall, C. & Rossman, G. (1995) *Designing Qualitative Research*, Sage Publications, Thousand Oaks, CA.

Mead, G.H. (1934) *Mind, Self and Society*, University of Chicago Press, Chicago.

Mihram, G.A. (1972) *Simulation: Statistical Foundations and Methodology*, Academic Press, New York.

Mohr, L.B. (1982) *Explaining Organizational Behavior: The Limits and Possibilities of Theory and Research*, Jossey-Bass, San Francisco.

Oakley, A. (1981) Interviewing women, a contradiction in terms. In: *Doing Feminist Research* (ed. H. Roberts), pp. 30–61. Heinemann, Oxford.

Pettigrew, A.M. (1997) What is processual analysis?, *Scandinavian Journal of Management*, **13**(4), 337–348.

Popper, K.R. (1989) *Conjectures and Refutations: The Growth of Scientific Knowledge*, Routledge, London.

Reber, A.S. (1995) *Dictionary of Psychology*, 2nd edn, Penguin, Harmondsworth.

Remenyi, D., Williams, B., Money, A. & Swartz, E. (1998) *Doing Research in Business and Management. An Introduction to Process and Method*, Sage, London.

Rooke, J., Seymour, D.E. & Fellows, R.F. (2003) The claims culture: a taxonomy of attitudes in the industry, *Construction Management and Economics*, **21**(2), 167–174.

Rooke, J., Seymour, D.E. & Fellows, R.F. (2004) Planning for claims: an ethnography of industry culture, *Construction Management and Economics*, **22**(6), 655–662.

SERC (1982) Specially promoted programme in construction management, *Science and Engineering Research Council Newsletter*, Summer, SERC.

Taha, H.A. (1971) *Operations Research*, Macmillan, Basingstoke.

Weber, M. (1978) *Economy and Society* (2 volumes), University of California Press, Berkeley, CA.

Zadeh, L.A. (1965) Fuzzy sets, *Information and Control*, **8**, 338–353.

Zimmermann, H.J. (2001) *Fuzzy Set Theory and Its Application*, Kluwer Academic Publishers, London.

Chapter 4

Approaches to Empirical Work

The objectives of this chapter are to:

- consider the role of 'experience';

- discuss research design;

- discuss qualitative and quantitative approaches;

- examine the requirements of *experimental design* (including experiments and quasi-experiments);

- examine case study research;

- examine the requirements of *modelling* and *simulation*;

Note: The word *experiment* includes *quasi-experiment* for the purpose of this chapter.

Role of experience

When does research begin?

It is all too common for people to believe that research has not really started until data collection has begun. That is wrong. A problem which occurs very frequently is that the data collection is begun prematurely – before the theory and literature has been reviewed and, in extreme cases, before the proposal has been finalised. It is hardly surprising that, in such cases, two significant problems arise concerning the data. The problems include collecting data which are not relevant to the research, and failing to collect data which are necessary. There may be further difficulties over the size and structure of the sources accessed to obtain the data. It is important to remember that it is difficult enough to collect data once; having to collect a second, supplementary set is compounding the difficulty geometrically. A researcher and, by implication, other researchers, will lose credibility by returning to respondents in order to remedy gaps in data collected. The target is to obtain high-quality data and only data which are relevant to the research being undertaken – get it right first time, you are unlikely to get a second chance.

What is experience?

Strictly, empirical work is concerned with knowledge gained from experimentation. Commonly much of that knowledge is gained through experience. Broadly, human knowledge exists in two main categories: express knowledge which can be stated/documented by its possessor ('I know what I know'); and tacit knowledge which can be demonstrated but is difficult to articulate, such as the attributes required to swim, or to ride a bicycle ('I can do that but I cannot explain fully how it is done'). So, knowledge is an outcome of the process of experience; the other outcome is expertise, which is a high level of skill in executing tasks (e.g. bricklaying, producing a project programme).

That leads to the question of 'what is experience'? Generally, experience is regarded as a form of human learning – learning by doing or by observation of doing by others. By definition, it involves observation, evaluation, memory and recall. All four activities include problems of selection and accuracy, so experience is unlikely to be totally reliable – observations, depend on perspectives and perceptions, some observations are missed, others are interpreted and understood incorrectly, memory can distort and recall may lead to omissions, i.e. memory is selective and deficient. Thirty years' experience may be of doing many new and different things over the period and so, such experience is cumulative; alternatively, it may involve doing the same things many times over (thirty repeats of one years' experience). In most cases, a combination of the two extremes is usual.

Usually, people are blamed more frequently than they are praised and remember blamings much more vividly. Not only does that demonstrate selective recall etc. but also has implications for behaviour both of the individual and of other people. 'Bosses' tend to blame more readily and more frequently than they praise. The consequence of such behaviour, the more vivid recall of blamings and the selectivity of memory, means that experience induces people to focus on avoidance of repeating mistakes. This may invoke conservative behaviour – perpetuation of the *status quo* in terms of performance etc.

A further consideration for 'experiential learning' is that it may lead to perpetuation of bad practices/mistakes. Accepting that the goal of any learning, research especially, is continuous improvement (of knowledge, skills, practices and, thence, performance), it is essential that real learning occurs via experience – that requires reflection, evaluating what has been done or observed in the context to theory and 'best practices'. Such evaluation is to ensure that the 'experiential learning' relates to improvements (if only through awareness of bad practices etc. and so, promoting avoidance of them in the future).

What is important for research is that there are problems of comprehensiveness and accuracy in relying on experiences – hence, the necessity to record all data accurately and speedily, irrespective of the research methods adopted; this necessity is well known and practised rigorously by those concerned with laboratory-based, experimental research and by those undertaking ethnography.

Research design

In research design, one has to decide the methodological approach in finding solutions/answers to the research problem or research questions (see Chapter 3, in particular, philosophy and methodology). It is about stating the way in which the researcher accomplishes the research objectives. In scientific research, one has to consider:

Types of research
  Examples: exploratory, descriptive, causal (explanatory, predictive)

Research approach
  Examples: quantitative, qualitative, triangulation

Empirical design
  Examples: between-subject design, within-subject design, longitudinal case study

Data collection methods
  Examples: survey, interviews,

Data analysis methods
  Examples: t-test, ANOVA.

Empirical design (how to collect and evaluate the data) is related to the overall research design. Sometimes, the research is carried out in stages; the research design takes the timing of the stages into account and the appropriate adoption of positivistic and phenomenological approaches in the various stages.

Empiricism and verification

Empiricism is a broad-based philosophical position grounded on the fundamental assumption that all knowledge comes from experience. Empiricism (in contrast to rationalism) advocates the collection and evaluation of data. An empirical test is the evaluation or assessment of a hypothesis or theory by appeal to data, facts and experimentation.

> *Example*
> Leary (2004, p. 9) gives the example that a behavioural researcher who is interested in the effects of exercise on stress would not simply chat with people who exercise about how much stress they feel – the researcher can design a controlled study in which people are assigned randomly to different exercise programmes then measure their stress using reliable and valid techniques – so that one may draw more confident conclusions on causality.

Rationalism is a philosophical perspective with the assumption that truth is to be ascertained through the use of reason (rational thought), e.g. Plato. Modern perspectives of rationalism are not entirely anti-empirical. Rational deductions are generally treated as susceptible to empirical demonstration and test. Hence, an empirical paper may report on findings based on data, facts and experiments, and a theoretical paper may report on the development in theoretical arguments and rational reasoning; however, these are not mutually exclusive.

The word empirical, has a number of meanings (Reber 1995): (1) relating to facts in general, (2) relating to experience in general, (3) descriptive of procedures carried out without explicit regard to any theory, (4) a general synonym of experimental, (5) descriptive of any procedures based upon factual evaluations, (6) pertaining to empiricism. All of these usages 'are based on or relate directly to data, to its collection, analysis or evolution. It is in this general sense that the term is most frequently used' (Reber 1995, p. 249).

The distinctions among the classifications of types of research are not absolute and a research project may involve more than one type of research design, e.g. exploratory research may be an initial step followed by descriptive or explanatory research. Some examples of types of research are given in Table 4.1.

While empiricism relies on observations to draw conclusions about the world, it is of paramount importance that conclusions should not be based on unfounded beliefs or assumptions. Hence, scientific observation is systematic, so that researchers can draw valid conclusions from the observations.

Table 4.1   Types of research.

Types of Research	Nature	Examples of Empirical Design	Common Data Collection Approach
Exploratory	To investigate phenomena and identify variables and generate hypotheses for further research	Case study Field study	
Explanatory	To explain causality	Multi-site case study Field study Ethnography Experiment Quasi-experiment	Participant observation In-depth interviews Survey questionnaire Document analysis Kinesics/proxemics
Descriptive	To document the phenomenon of interest	Field study Case study Ethnography	
Predictive	To predict outcomes and to forecast events and behaviours	Experiment Quasi-experiment	

> *Example*
> 'The scientists saw a flock of sheep standing in a field. One scientist said, "Look, all those sheep have just been shorn". The other scientist narrowed her eyes in thought, then replied, "Well, on the side facing us anyway".'   (Leary 2004, p. 9)
>
> 'That's almost a contradiction in terms; the face is usually on an end', said the third.

The research must be available for *verification*, i.e. the findings of one researcher can be observed, replicated and (tested) verified by others. Errors in methodology and interpretation can be discovered and corrected. Public verification requires that researchers report their methods and their findings to the scientific community, e.g. publishing journal articles or presenting in conferences. The *reliability* of a study refers to what happens if it is carried out again, i.e. if it is replicated. *Generalisability* means the ability to transfer a set of results from a particular group (sample) to apply to a much larger group (population) and this depends on good sampling. *Internal validity* is the degree to which a researcher draws accurate conclusions about the effects of an independent variable. *External validity* is the degree to which the results obtained in one study can be replicated or generalised to other samples, research settings and procedures. *Discriminant validity* documents the validity of a measure by showing that it does not correlate with measures of conceptually unrelated constructs. *Concurrent validity* is a form of criterion-related validity that reflects the extent to which a measure allows a researcher to distinguish between respondents at the time the measure is taken. Researchers often examine *construct validity* by calculating correlations between the measure of the construct and scores on other measures.

Researchers should report their methods in full detail, so that other researchers can attempt to repeat, or replicate, the research to allow other researchers to build on and extend the work.

Deduction and induction

Positivism tends to focus on facts, look for causality and reduce phenomena to their simplest elements for formulating (and testing) hypotheses; phenomenologism tends to focus on meanings, looking at what is happening in totality, developing ideas through induction and using small samples for in-depth investigation over time (see Easterby-Smith *et al.* 1994, p. 27 and Remenyi *et al.* 1998, p. 104).

Parsimony is important and it is necessary to theorise – to provide a theoretical explanation of observations by examining the processes of deduction and induction, importance of literature and adopting appropriate research design.

Phenomenology often relies on the induction process to generate the research question. Gould (1988, p. 22) argues that 'science, since people must do it, is a socially embedded activity. It progresses by hunch, vision and intuition ... The most creative theories are often imaginative visions imposed upon facts.' The deduction process looks for ideas in texts (journal articles, books etc.) and through communication with others (colleagues, experts etc.). However, whether the inductive or deductive process is adopted, it is fundamental that the conceptual framework be thoroughly developed, hence, the importance of literature and theorising. Even when non-positivists are using real-life problems as the inspiration for research, it is important that this is done with an understanding of the literature. Intuitive notions are, then, narrowed down into a researchable, informal hypothesis. The common methods employed in construction research in recent years are surveys (both in-depth and large scale), interviews, simulation and stochastic modelling, participant-observation and laboratory experiments. Some of the empirical designs are more reliant on the interpretivist approach than others (Table 4.2).

Case study

A case study is a detailed study of a single individual, group/organisation, or event/project. The data for case studies can come from a variety of sources, including observation, interviews, questionnaires, reports and

Table 4.2   Empirical design.

Empirical Design	Approach
Case studies	Have scope to be either positivist or interpretivist
In-depth survey by means of interviews	Mostly interpretivist
Large-scale survey by means of questionnaires	Positivistic with some room for interpretation
Simulation and stochastic modelling	Positivistic with some room for interpretation
Participant-observation	Strictly interpretivist
Laboratory experiments	Positivistic with some room for interpretation

Adapted from Remenyi *et al.* (1998).

archival records (such as minutes of meetings). The case study approach has at least four uses in construction management research:

(1) as a source of insights and ideas
(2) to describe phenomena
(3) project-biography
(4) illustrative anecdotes.

One important use of case studies is to provide a *source of insights and ideas* in the early stages of investigating a topic, e.g. studying a few key project participants in detail can provide a wealth of ideas for future investigation. Some *phenomena* do not occur frequently enough for the researcher to obtain a large number of participants displaying the phenomenon for study, e.g. specific types of construction accidents. Project-biography involves applying concepts and theories in an effort to understand the management of the project. *Project-biography* necessarily involves post-hoc explanations, e.g. project success factors. Even though interpretations of case study evidence are always open to debate, the systematic study of past projects adds a new dimension to knowledge. Researchers often use case studies as *illustrative anecdotes* to illustrate general principles to other researchers and to students. Supplementing hard empirical data with illustrative case studies may be valuable as they provide concrete, easy to remember examples of abstract concepts and processes.

Ethnography

Ethnography is an example of a methodology which departs from the positivist's quantitative research paradigm which is committed to establishing correlations between objectively defined variables as a basis for explanation. It is, essentially, the observer who stands at the heart of ethnography.

First, ethnographers study people in their natural settings, seeking to document that world in terms of the meanings and behaviour of the people in it. Second, ethnography does not follow the sequence of deductive theory testing; theory is often generated rather than solely tested – see grounded theory (Glaser & Strauss 1967). Third, the observer is the primary research instrument, accessing the field, establishing field relations, conducting and structuring observation and interviews.

Ethnocentrism refers to the practice of judging a different society by the standards and values of one's own (Seale 1998, p. 326). Phenomenocentrism is the tendency to accept one's own personal, immediate experience as revealing the true aspects of mind (Reber 1995). Ethnocentrism and phenomenocentrism are primary issues in the case of culture research when the researcher is alien to the culture being studied.

> *Example*
> An example is Geertz's (1973) interpretation of the Balinese cockfights – in which Geertz, through an intensive description of a cockfight, makes broader cultural interpretations and generalizations of reflexive interpretation (rather than making an objective description). Further deconstruction of ethnographic writing can be found in Rosaldo's (1989, p. 47) example of a breakfast scene. 'The bottom line seems to be that researchers should be aware of their rhetorical strategies ... It would be a disaster, in my view, if these insightful perspectives resulted in little more than a self-referential endo-professionalism, where research is reduced to endless textual deconstruction' (Back 1998, p. 292).

The Popperian argument is that validity involves confidence in our knowledge but not certainty. Thus, reality can be taken as independent of the claims researchers make about it. The production of truth rests on three things: the plausibility of the claim, given our existing knowledge; the credibility of the claim given the nature of the phenomena; and the circumstances of the research and the characteristics of the researcher (Walsh 1998, p. 232). Walsh (1998) suggests that validation of ethnographic research can take the forms of respondent validation and triangulation. However, respondents may not be privileged observers of their own actions or consciously aware of what they do and why. Triangulation involves comparison of data relating to the same phenomenon but derived from different phases of fieldwork, different points in time, accounts of different participants and using different methods of data collection – such as quantitative and qualitative. While various statistical methods are usually employed in quantitative research, content analysis, semiotics and discourse analysis are employed in qualitative research (e.g. Slater 1998; Tonkiss 1998; Silverman 1998).

> *Example*
> Rooke *et al.* (2003, 2004) employ ethnography, including participant observation, to explore various aspects of the, widely perceived, 'claims culture' of the construction industry in the UK.

Experiments and quasi-experiments

In experimental (and quasi-experimental) design, one looks at the relationships of variables – usually the independent, dependent, moderating, and intervening variables. An independent variable has at least two levels (or two conditions for the study). To ensure that the independent variable is strong enough to produce the hypothesised effects, researchers sometimes perform pilot tests.

The researcher often manipulates (varies) one or more independent variables to assess the effects on the dependent variable(s). In some research, one level of the independent variable involves the absence of the variable of interest. Participants who have a non-zero level of the independent variable compose the experimental group and those who receive a zero level of the independent variable make up the control group. Often, the researchers are interested in the effect of the independent variable on one or more dependent variables. A dependent variable is the response being measured in the study and the dependent variables typically involve either observations of actual behaviour or self-report measures (of participants' thoughts, feelings or behaviour) in recent construction management research.

Repeated measures design (within-subjects design) means each participant takes part in all conditions of the study. *Independent groups design* (between-subjects design) is where the same participant does not do all the conditions and different people are used for each condition. In between-subjects designs, researchers use simple or matched random assignments. In within-subjects or repeated measures designs, all participants serve in all conditions, thereby ensuring their equivalence.

> *Example*
> Organisational typology (independent variable) affects perceived bureaucracy (dependent variable).
>
> Condition 1: private organisations
>
> Condition 2: public organisations
>
> continued

> *Example   Continued*
> Between-subject design: data are collected from two groups of participants, those who have experience in private organisations and those who have experience in public organisations, i.e. different groups of participants giving responses to two different conditions.
>
> Within-subject design: data are collected from participants who have experience in both private and public organisations, i.e. the same group of participants giving responses to both set of conditions.

In studying the effect(s) of the independent variable on one or more dependent variables, the study could fail because the independent variable is not manipulated successfully. Hence, researchers often conduct pilot tests on a small number of participants before the full-scale study to see whether the effects of the independent variables are strong enough to be detected by the participants.

> *Example*
> In a pilot test to see whether monetary reward affects performance, a pilot test is not meant solely to reveal any effect of monetary reward on performance (which is the subject matter for the full-scale study) but to assess what might constitute levels (insignificant, satisfactory, or highly significant sums) of reward. Researchers can, then, be sure that the levels of the independent variable are sufficiently strong before the full-scale study is carried out to test its effect(s) on participants' performance.

Variance and errors

Variance is a numerical index of the variability in a set of data. *Confound variance* (systematic variance) is that portion of the variance in participants' scores that is due to extraneous factors that differ between the groups of participants and should be eliminated as far as possible. Experimental control refers to eliminating or holding constant extraneous factors that might affect the outcome of the study. Confounding occurs when something other than the independent variable differs in a systematic way and impacts on the results. The research is internally valid when it eliminates all potential sources of confound variance. When the study has internal validity, a researcher can confidently conclude that observed differences are due to variation in the independent variable.

Error variance (unsystematic variance) is the result of differences among participants, e.g. ability, personality, past history etc. Error variance is due to differences among participants within the group and should be reduced as far as possible, i.e. error variance is not due to the independent variable because all participants in a particular condition receive the same level of the independent variable, nor due to confounding variables because all participants within a group would experience any confound that exists. Error variance results from the uncontrolled and unidentified variables that affect participants' responses in the study, e.g. individual differences (ability, emotional states), etc. Error variance does not undermine the validity of the research, but researchers should try to minimise error variance, e.g. researchers should avoid differential treatment of the participants – try to treat/brief/inform all participants in precisely the same way.

A null hypothesis is the hypothesis that the independent variable will not have an effect. A *Type I error* occurs if one erroneously rejects the null hypothesis when it is true, i.e. concluding that an independent variable had an effect when, in fact, it did not. A *Type II error* occurs if one erroneously fails to reject the null hypothesis when it is false, i.e. concluding that the independent variable did not have an effect when, in fact, it did.

The limitations in the case study approach are, particularly, *observer biases* and difficulty in controlling *extraneous variables*. Observer bias occurs when the case study relies on the researcher's observations and interpretations where reliability and validity are difficult to determine. Case studies often deal with observation of isolated projects that occur in an uncontrolled environment and without comparison information; researchers are unable to assess the viability of alternative explanations of their observations. No matter how plausible the explanations offered for the individual's behaviours or for the effectiveness of a given treatment/method (e.g. site safety procedures), alternative explanations cannot be ruled out.

Qualitative approaches

When are qualitative approaches employed?

For a number of years, the scientific method, with an emphasis on positivism and quantitative studies, has been in the ascendant, with a result that research in disciplines which lie between the natural sciences and the social sciences, notably management of technology and engineering, has been drawn or pushed towards adoption of quantitative scientific method. However, quite recently, increasing recognition of the value and appropriateness of qualitative studies has emerged. This may, perhaps, be in acknowledgement of the

potential for such methodologies to get beneath the manifestations of problems and issues which are the subject of quantitative studies and, thereby,
to facilitate appreciation and understanding of basic causes and principles,
notably, behaviours.

Tesch (1991) identified three categories of approach to the analysis of
qualitative data:

- Language based – focuses on how language is used and what it means –
  e.g. conversation analysis, discourse analysis, ethnomethodology and
  symbolic interactionism. Understanding 'symbols' in the environment –
  language, gestures etc. and, hence, interpreting intent.
- Descriptive or interpretive – attempts to develop a coherent and comprehensive view of the subject material from the perspective of those who are
  being researched; the participators, respondents or subjects.
- Theory-building – seeks to develop theory out of the data collected
  during the study; *grounded-theory* is the best known example of this
  approach.

The approaches recognise that meaning is socially constructed, is negotiated
between people, and changes continuously over time. Therefore, it is important to examine and to take account of social interactions in the development
of theory and, wherever possible, to note the extent and direction(s) of the
dynamics of changes.

Oakley (1994) suggests that the word 'qualitative' is used to describe
research which emerges from observation of participants. She asserts that
such research has two sources:

- Social anthropology and
- Sociology.

Sociological studies often were conducted on westerners by westerners which
enabled the researcher, as a member of the population under study, to use
knowledge of the society to isolate themes and to prepare frameworks within
which study of a particular aspect could proceed. As anthropological studies were carried out by westerners on non-western societies, the work had
to be more 'open'. The researcher could not have valid preconceptions of the
society and so, not being part of it and having no initial understanding or
knowledge of it, could not isolate themes or provide a framework for restricting the scope of the study. Hence, all data had to be captured and examined
to enable hypotheses and theories to emerge.

While sociologists were able to employ questionnaires and interviews,
because they are influenced by the cultures in which they are devised and

conducted, the more basic, anthropological studies could not employ such techniques. All studies involving people are influenced by cultures; to a degree, that must include all studies.

> *Example*
> The topic of *indexicality* notes that a person's understanding of a term etc. is dependent upon cultural and contextual factors. Such a consideration is important when examining conditions of contract, perhaps for the purpose of researching disputes and claims in the construction industry (see, e.g., Clegg 1992). Clearly, it can be a very important and problematic consideration when drafting a contract for use internationally.

A consequence of considering the nature of the subject to be researched is that it may not be possible to isolate a particular, defined topic to study – what can be studied emerges from the research through what is observed. In such cases, it is not possible to develop a hypothesis to test. Furthermore, the aims and objectives are likely to be framed loosely and be quite 'open-ended'. Commonly, the subject of culture arises in debate over how members of an industry behave in various circumstances – whether concerning observations of behaviour or predictions. For example, it is said that the construction industry has 'a macho culture' and that it has 'a culture of conflict'. Usually, such expressions are the result of casual observations and are influenced by the values and experiences of the observer. In fact, very little research has been done to determine the cultures of the industry – the values and beliefs which govern people's behaviours.

> *Example*
> Consider researching the culture of engineers. As a non-engineer, and given appropriate definitions of terms, the research should first investigate codes of conduct etc. and formalisations of expected behaviours of engineers and then proceed to observe how they operate in order to devise hypotheses over cultural factors. Those hypotheses may be tested subsequently through case studies, interviews, questionnaires etc.

Of course, other different approaches could be employed. The findings may be common, different or a mix but, irrespective of that, their validities and applicabilities will vary, dependent upon how the studies have been carried out. Execution of such open-ended studies requires not only meticulous recording of all data, but constant scrutiny of the data to aid the recognition of themes – variables and patterns of relationships between them.

Development of theory from data

Developing theory from data requires much interaction between the researcher and the observed (persons); this interaction must be continuous over quite a long period, as well as exhaustive. For the researcher, much movement is required between data collection and data analysis to the extent that after many iterations the boundary between collection and analysis is likely to become fuzzy.

The strategies of the open-ended research approach are encapsulated in *grounded theory* – see Glaser and Strauss (1967) – which involves the discovery of theory from data. The technique involves the gathering of data from observations of the sample. Next, the researcher examines the data from the perspective of the issues to be investigated through the research, and identifies categories of the data. Further collection of data follows until, with continual examination of the data and the categories, the researcher is satisfied that the categories are suitable, that they are meaningful and important – that they are *saturated*. Saturation concerns production of categories for data allocation and collection of data. Saturation occurs when further assignment of categories is not meaningful because the categories developed are complete for the purposes of the research; in the case of data collection, saturation occurs where collection of further data provides no additional information about the attributes of any category.

Morse (1994) suggests that three phases are involved in research. *Comprehension* requires development of an indicative model from any theory or literature available (this may not be possible – in some cases of fundamental research nothing relevant may have been published), followed by collection of data. Comprehension is achieved when enough data have been collected (by observation/unstructured interviews) from the full spectrum of participants' perspectives to provide in-depth understanding of the subject matter of the research (similar to *saturation*).

The second phase is *synthesis*. Initial analyses of data collected suggest further aspects to be researched. Those further collections and analyses of data continue until the third phase, *saturation*, occurs; when further data and their analysis no longer provide additional insights or indications of further aspects meriting investigation, and so no change in the understanding that has been developed.

Further work can be undertaken to investigate generalisation of the categories and links between them. Such work will employ hypotheses to be tested by additional field work with a new or extended sample and may employ the technique of *analytic induction*. Analytic induction is a step-by-step process of iteration and evaluation. Initially, the issue is defined (perhaps only approximately), instances of the issue are examined and potential explanations and relationships are developed. Further instances or samples are investigated to analyse how well the hypothesised explanations and relationships apply. Such iterations continue until the hypotheses suit what is found in the data at a suitable level of statistical significance in appropriate cases.

In using grounded theory and analytic induction, Strauss (1987) emphasises that researchers must be, '... fully aware of themselves as instruments for developing that grounded theory.' The statement means that the researcher must be rigorous and highly objective in analysing the data to yield the categories and, thence, theory. In instances where hypotheses have been developed for testing, any desires to support or refute the hypothesis by the researcher must be ignored. In the research process, to ensure accuracy and validity, the research must *avoid bias*.

In carrying out studies which require the development of theory from data and the subsequent testing of the theory, where data collection, analysis and development of theory proceed together iteratively, Schatzman and Strauss (1973) advocate segregation of the researcher's field notes into:

- Observational Notes (ON),
- Theoretical Notes (TN) and
- Methodological Notes (MN).

Observational notes (ON) concern the recording of '... events experienced principally through watching and listening. They contain as little interpretation as possible and are as reliable as the observer can construct them.' Theoretical notes (TN) are 'self-conscious, controlled attempts to derive meaning from any one of several observation notes'. Methodological notes (MN) concern how the field work is carried out, and record any necessary changes, the reasons for such changes and when the changes occurred. Irrespective of the research methodology adopted for any project, taking detailed field (laboratory) notes is vital. The categorisation advocated by Schatzman and Strauss is appropriate for any research project, whether using qualitative or quantitative methods.

Observers have to select what is recorded – that may introduce bias. The problem is accentuated in participative (participant) observation as the researcher is executing the two functions of participating in the activity and observing (plus recording and, possibly, analysing) it concurrently. Pre-designed, structured forms for recording data help overcome some problems (notably bias) but may, of course, lead to important, but not predetermined observations being omitted.

Much qualitative research concerns the generation of concepts through the researcher getting immersed in the data collected in order to discover any patterns. In so doing, it is essential to be sensitive in order to detect inconsistencies and to be aware of the potential for different views to be expressed and for alternative categorisations and explanations to be valid. The researcher must be aware of her/his own preconditioning and views – potential bias.

A particular problem in qualitative research is 'ethnocentrism' – understanding others and/or interpreting their behaviour on the basis of one's own values. Researchers should endeavour to be 'value free' in order to be able to interpret others' behaviour from *their* perspective, which is, of course, extremely difficult.

> '… qualitative data often provide a good understanding of … the "why" of what is happening. This is crucial to the establishment of internal validity … [and to] … discover underlying theoretical reasons for why the relationship exists.'   (Eisenhardt 1989)

Analysis of data

Hammersley and Atkinson (1983) are amongst a number of authors who consider the construction of *typologies* and *taxonomies*, which are categories and groups within the categories, to be important elements of analyses. The researcher should seek to establish categories, subgroups and relationships between them from the data collected. Such categorisation of data will reduce the number of potential variables, thereby making the data more manageable and 'visible' to assist the detection of patterns and possible dependencies, also called causalities. Clearly, in such qualitative research, much analysis is carried out by the researcher during the period of collecting data in the field.

Bogdan and Biklen (1982) differentiate between analysis carried out in the field during the period of collecting data and analysis carried out after the data collection has been completed. They assert that the researcher needs to be engaged in preliminary analyses constantly during data collection whilst post-collection analyses concern developing a system of coding the data primarily.

Charmaz (1983) believes that, 'qualitative coding is not the same as quantitative coding. … Quantitative coding requires preconceived, logically deduced codes into which the data are placed. Qualitative coding … means creating categories from interpretation of the data.' The belief implies that qualitative coding is more flexible, as categories are created to suit the data from the data collected, whereas quantitative coding may require data to be force-fitted into the preselected categories.

Harris and Ogbonna (2002) summarise three aspects of coding, which are important for analysing qualitative data, especially in application of grounded theory. 'Open coding involves the line-by-line analysis of text for the occurrence of categories and the subsequent deconstruction of data into emerging categories. During axial coding, these categories are reviewed and re-sorted, leading to the emergence of sub-categories linkages and relationships.... . Selective coding involves the evaluation of selected data that appears particularly relevant in refining the emerging interpretation.'

It is likely to be a poor idea to have a 'miscellaneous' category because, unless the data are allocated to the pre-selected codes as far as possible, albeit with effort to make the data fit into the categories, any difficulty in allocating an item of data will result in its entering the miscellaneous category. The consequence may be that 'miscellaneous' is the largest of all the categories, which may render analyses difficult, if not meaningless!

Quantitative approaches

When are quantitative approaches employed?

Essentially, quantitative approaches involve making measurements in collecting data. The approach is built upon previous work which has developed principles, laws and theories to help to decide the data requirements of the particular research project. Two major questions are: what is to be measured; and how should those measurements be made? One aspect which may be overlooked is the scale of measurement (see below and Chapters 6 and 7); the nature of the scale selected is very important – for ease of collecting data, for accuracy considerations and, perhaps most importantly, for determination of what analyses can be carried out validly.

The answers to the questions, above, are derived from examination of the theory and previous research findings together with the aim and objectives of the research to be carried out; in particular, the hypothesised relationships in the research model. Therefore, the coding framework is, as noted above, already in place.

In allocating data to categories in a database by means of coding exercises, Fine (1975) demonstrated some significant facets which are often overlooked. It seems to be a common belief that the more detailed or complex the database is, the more useful it will be. The main assumption is that of accuracy in allocation – that in using the coding system, the data are allocated accurately.

Example
Fine (1975) conducted an experiment in which construction 'cost accountants' were asked to allocate items to a cost database. The results were:

30 categories in the database, 98% items allocated accurately
200 categories in the database, 50% items allocated accurately
2000 categories in the database, 2% items allocated accurately.

Of course, the significance of misallocations can vary enormously depending on the nature of the items, the database, the nature of misallocations and the use of the database.

Given the early stages of quantitative methods to arrive at the field-work stage of the research, there is the requirement of a considerable amount of pre-conception in deciding what data are to be collected, how they will be collected and what analyses will be done. Without thorough study of the underpinning theory and literature, important factors are likely to be missed thereby, at the very least, reducing the validity of outputs from the research as well as causing difficulties in executing the work.

Sources of data

Ideally, the researcher and the existence of the research will have no influence on the data collected. However, that is known to be untrue and so the pragmatic objective is to minimise the impacts. Such minimisation is sought by using objective methods designed to remove as much bias as possible and to conduct the research in the most unobtrusive way, whilst retaining good-will of the collaborators and subjects of study – essential in studies of people and their behaviours.

Throughout quantitative studies, and scientific method, a major objective is that the research is 'value-free'; that the work is unaffected by the beliefs and values of the researcher(s) – it is objective. In conducting quantitative research, three main approaches are employed: asking questions of respondents by questionnaires and interviews; carrying out experiments; and 'desk research' using data collected by others. Using data collected by others, who may have collected it for a variety of other purposes, can be problematic, as the data, sampling etc. have not been tailored to the particular research project in question. However, it can be very helpful to use data collected already by others – it saves time, can be cheap and, for studies such as macroeconomics, can be the only viable way of obtaining the data required.

Clearly, it is essential to investigate the nature of the data and collection mechanisms in order to be aware of the limitations of the data and their validity, notably comparability – for instance, the basis for producing unemployment statistics in the UK has changed many times. For longitudinal studies, the difficulties of comparing like with like may be great.

In executing experiments, results are sought by effecting incremental changes in the independent variable and measuring the effect, if any, on the dependent variable, whilst holding all else constant. Using the experimental style in a social context produces problems far in excess of those encountered in a science research laboratory; society is dynamic, the number of variables operating is vast and having to carry out the research in the real social context prevents control over the variables (see discussions of *quasi experiments*). Hence, it is highly unlikely that only one variable will change during the study – the practical approach is to restrict the impact of environmental variables as far as possible.

Usually, the sample of people on whom the experiment is to be conducted is split into two groups which are matched as far as possible in all respects. The independent variable is changed for one group but is unaltered for the other, known as the *control group*. Having matched the samples in the two groups and holding constant all other variables which, in practical terms, act identically on both groups, it is therefore valid to assume that changing the independent variable will yield identical consequences with respect to the dependent variable. Hence, the differences measured between the two groups with respect to the dependent variable over the course of the experiment are due to the measured changes made to the independent variable. Thus, cause and effect, in direction and magnitude, are established. (The stages in experimental design are shown in Fig. 1.3, p. 12.) Clearly, very large samples of people are necessary for experimental techniques to be used in investigating behaviours, safety of medicines etc.

Experimental design in human behaviour involves developing strategies for executing scientific inquiry to enable the researcher to make observations and interpret the results. There are two important aspects to consider in formulating experimental design:

- unit of analysis
- time dimension.

It is critically important to identify the unit of analysis, such as the individual or the group, accurately. Failure to do so may result in two errors of logic: the ecological fallacy; and reductionism. The ecological fallacy involves gathering data on one unit of analysis but making assertions regarding another. The most common instance of the ecological fallacy is making inferences about individuals from statistics of the entire population – i.e. assuming everyone is 'typical'; stereotyping! The fallacy may operate in reverse also by attributing the behaviour of an individual to the entire population or group to which that person belongs.

Reductionism has a number of forms – ontological, methodological, theoretical, etc. In all of those, the basic concept is that a complex system can be understood best from examination of its constituents and the interactions between them. That perspective seems to ignore potential synergy. Holism incorporates synergy in positing that complex systems may have properties beyond the mere sum of their individual components. Reductionism also refers to an over-strict limitation on the kinds of concepts and variables to be considered in understanding a social phenomenon. Both errors involve misuse of the unit of analysis.

Another aspect of experimental design is the time dimension, as exemplified in *cross-sectional studies*, which are observations at one point in time, and *longitudinal studies*, in which observations are made at multiple, usually pre-selected, time points. The time intervals between observations may be equal over the investigation period or, especially if archival data are being collected, structured to capture the effects of important events. The three types of longitudinal studies are: *trend studies*, those that examine changes within some general population over time; *cohort studies*, those that examine more specific subpopulations as they change over time; and *panel studies*, those that examine the same set of people over time (Babbie 1992). Generally, longitudinal studies are superior to cross-sectional studies for making causal assertions, although many longitudinal studies are, in fact series of cross-sectional studies.

Longitudinal studies which employ a series of cross-sectional studies are one form of *repeat measures design*. Other projects may use repeat measures design to examine effects of different 'treatments' or to determine temporal consistency. Repeat measure designs may be subject to order of presentation bias under which memory of previous responses impacts the instant response, hence, for each set of collecting the responses, a freshly randomised sequence of questions (or other response-generating items) should be used. Further, time between sets of responses should be sufficient to allow respondents memories to fade. If the purpose is to examine the effects of different treatment conditions, a further approach is to use different sequences of treatment for each group of respondents to achieve 'counterbalancing' of any biases.

In formulating a good experimental design, it is important to consider:

- How to vary an independent variable to assess its effects on the dependent variable.
- How to collect data or, in the case of social behavioural research, how to assign subjects to the various experimental conditions.
- How to control extraneous variables that may influence the dependent variable.

In experimental design, the researcher is interested in the effect of the independent variable on one or more dependent variables. A dependent variable is the response being measured in the study. In a between-subjects design, subjects are randomly assigned to experimental conditions, and data can be collected from various groups of subjects. In within-subjects design, all subjects serve in all experimental conditions.

> *Example*
>
> Leary (1991, 2004) gives the example of testing the effect of caffeine, the independent variable, on subjects' memories, the dependent variable. The independent variable can be varied in terms of quantitative differences, such as different quantities of caffeine given to the subject, to test its effect on the dependent variable, in this case, the subjects' memories of a list of words. In other circumstances, the independent variable could be varied in terms of qualitative differences, such as the colour of a cup of coffee. One level of the independent variable can involve the absence of the variable of interest. Subjects who receive a non-zero level of the independent variable compose the experimental groups and those who receive a zero level (say, no caffeine) of the independent variable constitute the control group. Although control groups are useful in many experimental designs, they are not always necessary.
>
> In a between-subjects design, each group is given a different level of caffeine consumption; the researcher must be able to assume that subjects in the various experimental groups did not differ from one another before the experiment began. This ensures that, on average, subjects observed under the various conditions are equivalent. Alternatively, in within-subjects designs, a single group of subjects is given all levels of caffeine consumption. As such, the researcher is testing the differences in behaviour across conditions within a single group of subjects.

Within-subjects design is better than between-subjects design for detecting effects of the independent variable. This is because the subjects in all experimental conditions are identical in every way so that none of the observed differences in responses to the various conditions can be due to pre-existing differences between subjects in the groups. Since repeated measures are taken on every subject, it is easier to detect the effects of the independent variable on each subject.

Experimental control

Experimental control is essential; it refers to eliminating, or holding constant, extraneous factors that might affect the outcome of the study. (See also: validities – Chapter 5; obtaining data – Chapter 6.)

Nothing other than the independent variable should vary systematically across conditions, otherwise, confounding occurs. Confounding destroys the internal validity of the experiment, which is a very serious flaw. Researchers should also try to minimise sources of error variance which is produced by unsystematic differences between subjects within experimental conditions. Although error variance does not undermine the validity of an experiment, it makes it more difficult to detect the effects of the independent variable. However, attempts to reduce the error variance in an experiment often reduce the study's external validity; the degree to which the results can be generalised.

In most instances, researchers do not have the necessary control over the environment to structure the research setting, and *quasi-experimentation* results (Condray 1986). Quasi-experimentation is a pragmatic approach which attempts to collect the most meaningful data under circumstances that are often less than ideal. Rather than adhering to just one particular experimental design, researchers may use whatever procedures are available to devise a reasonable test of the research hypothesis to maximise reliability and validity. Most often, given the absence of random assignment of subjects, as in between-subject design, simply showing that a particular quasi-independent variable was associated with changes in the dependent variable may not be convincing enough to show that the quasi-independent variable caused the dependent variable to change. The researcher may also have to demonstrate that the quasi-independent variable was associated with changes in other variables assumed to mediate the change in the dependent variable. By looking at other additional quasi-independent variables, comparison groups and dependent measures, researchers increase their confidence in the inferences they draw about the causal link between the quasi-independent and dependent variables.

Experimental design (including experiments and quasi-experiments)

Experiments and quasi-experiments

An experiment is an activity or *process*, a combination of activities, which produces *events*, possible outcomes. Usually, in scientific contexts, experiments are devised and conducted as tests to investigate any relationship(s) between the activities carried out and the resultant outcomes. Tossing a coin a number of times could be used as an experiment to test for bias in the coin; likewise, the throwing of a die. Hicks (1982) defines an experiment as a 'study in which certain independent variables are manipulated, their effect on one or more dependent variables is determined and the levels of these independent variables are assigned at random to the experimental units in the study'.

Ideally, variables should be isolated through the design of an experiment such that only one of the, possibly, very many independent variables' values is changed and the consequences on the isolated single dependent variable is monitored and measured accurately (akin to sensitivity analysis). Hicks' definition raises the issue of the way in which the independent variable is 'manipulated'; although random variation is one approach, commonly, particular values within a 'range of interest' are assigned to the independent variable. This method provides practicality, but also some restriction on the inferences which can be drawn from the results.

In social investigations, including construction management and construction project-based 'experiments', it is neither practical nor possible to allow only one independent variable to alter in value, nor is it possible to isolate individual dependent variables on most occasions, hence the usual approach to experimental design is to devise a study in which the main independent variables, except the one of interest, are held *approximately* constant and the consequences for the major dependent variable are measured. Such approaches are called quasi-experiments. A common approach is to undertake comparative studies on similar projects executed at about the same time by similar firms employing similar organisational arrangements. Such a study could investigate the impact of different management styles of project managers on project management performance, as measured in terms of time, cost, quality, etc. Variables of location, weather etc. as well as (preferably, minor) differences between the 'common independent variables' (environmental variables which, through the research/experiment design, are held *approximately* constant) should be acknowledged in the evaluation of results.

An experiment is designed and it occurs in the future. However, there are many instances where analysis is required of data which have been collected in the past; such an approach cannot be an experiment but is known as *ex-post-facto* research. Ex-post-facto research is very useful in many contexts, such as building economic models from which forecasts can be made. In testing such models, it is essential to ensure that the data used for the test have not been used in the model-building (see sections on modelling and simulation).

Commonly, it is believed that experiments and their designs must begin with a statement of the problem or issue to be investigated. However, that requirement does not apply to research and, further, is not the real initiation of research; that initiation is recognition that a problem/issue may exist, and which gives rise to the question, 'what is the research intended to find out?'. Given that the question is asked, it must be answered as precisely as possible in order that a statement of the intended investigation can be made, noting requirements, parameters and limitations. In this way, the most appropriate methods by which to carry out the research can be determined. Thus, given resolution of the preliminary issues, the devising and design of an experiment begins with, and is driven by, the statement of the problem.

Variables

A variable is a quantity, the size of which may change – so, the duration of a construction project is a variable (until the construction has been completed). Variables are distinct from constructs – a variable is a single, definite phenomenon such as project capital cost; a construct is a combination of variables to yield a conglomerate phenomenon such as project quality. Variables should not be confused with target values or realised values which are particular quantifications of a variable. Variables can be grouped into five major categories: dependent, independent, moderating, mediating and intervening. Independent variables, for design of the research/experiment may be divided into the independent variable(s) which is to be the subject of measured changes to examine the effects on the dependent variable, and environmental variables which the design seeks to isolate from affecting the dependent variable during the research/experiment (by holding their values constant etc.).

A dependent variable is the subject of the research, experiment or test (the output of the system); its attributes (quantity, quality) are affected by (depend upon) the attributes of the other variables with which it is associated, especially the independent variable. Dependent variables are also known as response variables, responding variables, explained variables, criterion variables or regressands.

An independent variable is a phenomenon, the attributes of which are controlled, selected and (possibly only) measured by the researcher so that any corresponding attributes of the dependent variable can be examined; the relationship between the independent variable and the dependent variable are examined for nature, size and, most especially, for causality. Independent variables are also known as predictor variables, controlled variables, manipulated variables, explanatory variables or regressors.

'A moderator is a qualitative (e.g., sex, race, class) or quantitative (e.g., level of reward) variable that affects the direction and/or strength of the relation between an independent or predictor variable and a dependent or criterion variable.' (Baron & Kenny 1986)

'A given variable may be said to function as a mediator to the extent that it accounts for the relation between the predictor and the criterion.' (Baron & Kenny 1986) (see Fig. 4.1)

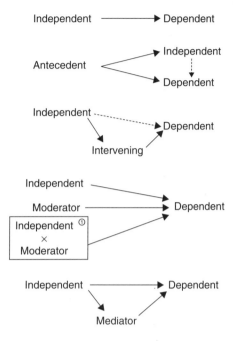

Note: ① This product of the two variables is a construct.

Fig. 4.1   Types of variables and relationships between them. (See also Baron & Kenny 1986; Drenth 1998 for more details.)

'An intervening variable is a hypothetical concept that attempts to explain relationships between variables, and especially the relationships between independent variables and dependent variables. It is often distinguished from a hypothetical construct in that it has no properties other than those observed in empirical research. That is, it is simply a summary of the relationships observed between independent and dependent variables. For example, *hunger* is a hypothetical internal state which has been used to explain the relationships between independent variables such as length of time without food, or amount of food consumed, and dependent variables which are measures of eating.'   (Wikipedia 2007)

In common with any information production, the objective is to support and facilitate decision making. In the case of experimental research, the decision making concerns inference about the relationships investigated. Hence, the variables must be identified, together with appropriate definitions and measures; assumptions should be explicit so that appropriate hypotheses for testing by the experimentation can be formulated expressly. Commonly, such hypotheses concern relationships between independent and dependent variables, to assist examinations of strength and direction of the relationships and, in view of theory and literature, causality.

Clearly, measurement of the variables is crucial. Experimental design considers the degree of accuracy which can be achieved and the method of achievement. Given that no measurement can be 100% accurate, the criterion is to obtain sufficient accuracy for the purpose of the experiment; if alternative experimental or measurement techniques are available, the most accurate means, subject to the pertinent constraints, should be selected. For example, in forecasting accuracy of an experiment, the probability of errors etc. should be considered.

A major consideration in designing an experiment is the method used to change the independent variable in order that any consequential changes in the dependent variable can be measured. Three main approaches to effecting the changes in the independent variable are employed:

- *Randomised change* – of the independent variables and/or their values – perhaps by use of some random number generator to determine the values to be employed, within or without limits to the 'range of interest'. Randomisation allows the experiment to be conducted, results produced and conclusions drawn, using the common assumption of independence of errors in measurement; this is usual in much statistical analysis – randomisation validates the assumption. Further, randomisation supports the assumption of 'averaging out' the effects of uncontrolled independent or intervening variables. Such averaging removes much of the effects of the uncontrolled variables but does *not* do so totally – those variables increase the variance of the measured values of the dependent variable. Randomisation is helpful in eliminating bias; it ensures that no variables, or their possible values, are favoured or disregarded as there is no 'systematic' selection of either variables or their values. Further, it ensures independence between observations and measurements, a requirement for validity of significance tests and interval estimates.
- *Selected ranges of variables* – both in terms of the identities of independent variables and the ranges of the values which they may assume. For such experiments, the main independent variables must be identified by scrutiny of theory and literature etc. The variables can be quantitative and/or qualitative and their values selected or random. Use of extreme values of the variables should result in maximum effects on the dependent variables and, hence, both the identification of the range of consequences which might result, and a chance to focus on the most likely outcomes.
- The most restrictive but, often, the most convenient/appropriate approach, is to *control the independent variables rigidly* – the variables are determined and values assigned over the duration of the experiment (as in Boyle's Law experiments etc.). Strictly, the inferences which may be drawn from such experiments are valid for the fixed experimental conditions only.

Given more than one independent variable of interest, a factorial experimental design results. If there are two independent variables, α, β, where α may assume (for the purposes of the experiment) 5 values and β may assume 3 values, the result is a 5×3 matrix such that 15 combinations of the values of the independent variables must be investigated to determine the effect of each value combination on the dependent variable.

Replication

A universal, desirable feature of research, notably experiments, is replication – hence, it is essential to make meticulous notes of record of *all* events in detailing the conduct of an experiment. Good laboratory practice involves detailed and precise recording of all occurrences during the conduct of an experiment for subsequent scrutiny. Without complete detail of experiments, replication is not possible. As replication facilitates increased numbers of observations and measurements of the variables under *identical treatments*, it assists provision of an estimate of experimental error, and identification and quantification of sources of error. Further, it should lead to reduction in standard errors, thereby increasing precision:

$$S_{\bar{y}} = \sqrt{\frac{s^2}{n}}$$

where: $S_{\bar{y}}$ = standard error of the mean (of the dependent variable, y)
s^2 = sample variance
n = number of observations.

The result is that replication assists inference; as replication increases, so a wider variety of situations and variables etc. can be subject to the experiment, thereby yielding a greater range of conditions to which the results apply, and so, the inference base is broadened.

Petersen (1985) suggests that, due to the nature of standard errors, the accuracy of experimental results can be improved by:

- Increasing the size of the experiment by
  - replication,
  - incorporating more 'treatments' (values of independent variables).
- Refining the experimental method or technique to achieve reduction of experimental error through reducing the sample variance (s^2).
- Measuring a *concomitant variable*, another, associated independent variable, to facilitate *covariance analysis*, the variance of the combination of the variables, this may yield reduced experimental error.

In considering factorial experiments, described below, Petersen notes the following, additional considerations:

- Increasing the size of the experiment (as above) can be self-defeating if it requires the incorporation of more heterogeneous experimental units.
- Accuracy can be increased through selection of treatments such that the factorial combinations include hidden replication for some comparisons.
- Arranging experimental units into homogeneous groups can remove differences between groups. This reduces sample variance (s^2) and hence, experimental error.

In examining the results of any experiment, it is likely to be helpful to carry out an analysis of variance (often identified as ANOVA or, for multivariate analyses, MANOVA, in computer statistics programs). Analysis of variance is a systematic approach which identifies the constituents of the total variation, and thereby apportions the total variation to the contributing sources. Such analysis can be very helpful in refining experimental designs.

Between-subjects design (simple randomised experiments)

This is the basic design in which values of variables are allocated at random. Much of statistics employs the assumption of such randomness and the statistical analysis tends to be quite straightforward. Randomised design offers greatest flexibility; however, its precision can be low, especially if experimental units are not uniform (Petersen 1985).

Between-subjects design (matched randomised groups)

The experimental groups or blocks are composed of units which are as near as possible homogeneous. This may be achieved by random allocation of units to groups or by precise design allocation of the units to groups. The aim is to avoid differences, especially 'systematic' differences, between the groups, thereby increasing precision by eliminating inter-group (between-group) differences from experimental error. Thus, to be effective, the intra-group (within-group) variance must be much smaller than the variance over the entire set of units. Hence, Petersen notes that:

- the size of each group should be as small as possible because, usually, precision decreases as the size of the group increases.
- If there are no obvious criteria for designing the group, 'square' group designs are most appropriate.

- If 'gradients' apply to groups (e.g. slope, age, strength), groups should be designed to be narrow and long (rectangular and perpendicular to the gradient).

Given Z treatments, each replicated n times, results in the need for nZ experimental units. The units should be placed into n groups, each of Z units, in a way so that the groups are as similar as possible. Then, the treatments are assigned to the units randomly such that each treatment occurs only once within each group. A notable problem in group designs is that missing data can cause problems for analysis.

The use of a randomised group design can increase the information obtained from an experiment, as the groups can be at different locations and the individual elements of the experiment can be carried out at different times. Thus, sampling can occur over a wider variety of circumstances. The aim is to separate the effects of the treatments from uncontrolled variations amongst the experimental units or groups. The treatments, units and their grouping and observations and measurements should be designed such that:

- Experimental units which are subject to different treatments should differ in no systematic way from each other (i.e. they should be unbiased; differences should be random, if any, and small).
- Experimental error should be minimised and achieved by use of as few experimental units in each group as possible.
- Conclusions should have maximum validity (concerning breadth and depth).
- Experimental technique should be as simple as possible, commensurate with the objectives (i.e. *parsimonious*).
- Statistical analyses of the results should not require the use of assumptions which are too restrictive or are inappropriate in the context of the objectives and the circumstances of the experiment.

Within-subject design (repeated measure design)

In within-subjects design, the researcher is interested in differences in behaviour across conditions within a single group of participants, i.e. a single group of participants serves in all conditions (all levels of the independent variable). For instance, the independent variable being organisational typology (private organisations vs. public organisations), the data are collected from participants who have experience in both private and public organisations, i.e. the same group of participants giving response to both sets of conditions.

Within-subjects design eliminates the need for random assignment as every participant is tested under every level of the independent variable – none of the differences in responses to the various conditions can be due to pre-existing differences between participants in the groups – it is easier to detect the effects of the independent variable on the dependent variable (e.g. participants' perception of bureaucracy in public vs. private organisations). However, carryover effects may occur when the effects of one level of the independent variable are still present when another level of the independent variable is introduced, e.g. the participant cannot be working in both public and private organisations at the same time, so the effect of his/her last employment (say, in a public organisation) may be carried over to the current employment.

Factorial experiments

Factorial experiments can be considered as constrained instances of randomised groups. For convenience, the main independent variables are identified and assigned values to be considered towards the extremes of their practical or usual range. Such an approach limits the number of combinations of continuous variables to be analysed. This is due to 'central limit' type of effects and the opportunity to interpolate between the results of the experiments so undertaken.

Example

To consider ready-mixed concrete from two suppliers, variables such as water–cement ratios, maximum aggregate size and amount of plasticiser could be assigned high and low values to yield a $2 \times 2 \times 2 \times 2 = 16$ 'cell' experiment, i.e. 16 possible combinations of factors affecting the strength achieved.

Note: Increasing the number of independent variables to be examined would extend the number of experiments considerably. However, the technique of 'Latin Squares' can be employed to restrict the number required and to maintain significance of and confidence in the results (see Levin & Rubin 1991, pp. 286–287).

Case study research

Case study approaches facilitate in-depth investigation of particular instances of a phenomenon. Those instances may be selected in a number of ways – to be representative of general cases (typical; selection of cases is akin to statistical sampling), bespoke case(s), random, etc. Normally, and because only a small number of cases are studied, but the studies are in-depth, the purpose is to secure theoretical validity (as for experiments), rather than the (more common) statistical validity required of surveys. Harris and Ogbonna (2002) note that '... case studies prove valuable in situations where existing knowledge is limited ...' and continue '... within business research, a case study is a description of a situation which is sensitive to the context in which the research occurs ...'.

In case study research, which investigates phenomena within context, often the contextual variables are so numerous and qualitatively different that no single survey or data collection approach can be appropriately used to collect information about these variables. Hence, Yin (1993, p. 2) contrasts case study design with experimental design where the focus is on testing one or two specific variables whilst others are 'controlled out' or 'kept constant' and describes case study research as 'an empirical enquiry in which the number of variables exceeds the number of data points'.

Pettigrew (1997) notes that the important choices for a researcher include '... the primary unit of analysis ... how context is to be defined and operationalised ... [and the] ... time frame of the study'.

The emulation of the scientific approach (logical positivism) in case study research, i.e. the clear steps of developing research questions, literature review to develop hypotheses, finalise methodology and research design, collecting and analysing data to draw conclusion, is emphasised by certain researchers, e.g. 'My only claim that case studies that follow procedures from, "normal" science are likely to be of higher quality than case studies that do not' (Yin 1993, p. xvi). Taking this approach, the rigour of case study research is judged by the same criteria of internal validity, construct validity, external validity and reliability as in other forms of scientific research.

Case studies are also used in ethnographic research. However, ethnographic research does not emulate the traditional paradigm of positivist empirical science which assumes a single objective reality that can be repeatedly replicated, but is guided by the assumption of multiple realities that are socially constructed rather than the belief that there is a single objective reality which is independent of human observers. Participant observation is the major data collection technique in ethnographic research (usually supplemented with interviews, archival searches etc.) since the researcher cannot maintain an objective distance from the phenomenon being studied; indeed, ethnography requires the researcher to become immersed in the life of the social group under study. Van Maanen (1979) cautions that 'The results of ethnographic study are thus mediated several times over – first, by the fieldworker's own standards of relevance as to what is and what is not worthy of observation; second, by the historically situated questions that are put to the people in the setting; third, by the self-reflection demanded of an informant; and fourth, by the intentional and unintentional ways the produced data are misleading'.

> *Example*
> Rooke *et al.* (2004) employed ethnography to study culture within construction organisations and gained important insights into the roles played by claims in the bidding process, how different groups of people regard each other on construction projects and how claims are pursued opportunistically.

Grounded theory, which is directed at theory building rather than theory testing, also uses case studies. A grounded theory is one that is inductively derived from the study of the phenomenon it represents (Strauss & Corbin 1990) and seeks to avoid premature use of theory or prior conceptual categories (Glaser & Strauss 1967). Grounded theory identifies emergent categories from empirical data by using qualitative data analysis methods but the data do not have to be field-based, e.g. documents from various (library) sources.

Yin (1993) asserts that case study research can be based on a (2 × 3) typology design, i.e. single- or multiple-cases mapped with exploratory, descriptive or explanatory study. Whilst a single-case study needs only to focus on one case, in multiple-case studies, cases should be selected so that they are replicating each other – either exact (direct) replications or predictably different (systematic) replications. Also, '... case studies can employ an embedded design, that is, multiple levels of analysis within a single study ...' (Eisenhardt 1989).

> *Example*
> Turner (1994) reports on research into the patterns of crisis behaviour discernible during a tragic and disastrous fire using grounded theory and case studies.
> Hughes (1994) reports on research utilising participant observation in studying (step-) family relationships.

Descriptive case study is aimed at systematically identifying and recording a certain phenomenon or process (e.g. see cases in Luthans 1992). It is not directly aimed at testing a theory or hypothesis but at recording an object of study. Through case studies, one tries to find new theoretical interpretations or to gain more in-depth knowledge pertaining to existing theoretical insights. Exploratory case study is theory-driven as 'Theory is a guide to tell you where to look for what you want to observe' (Runkel & McGrath 1972, p. 3). A large-scale research project is often preceded by a 'pilot' study which aims at generating hypotheses, in which case it is often termed an exploratory study. Explanatory research aims at hypotheses testing which, usually, has a causal explanatory character (based on a probabilistic relation) allowing a conclusion to be logically inferred, e.g. high levels of job satisfaction lead to low absenteeism etc. A large number of case studies fall between the two extremes of exploratory and explanatory research, e.g. field studies into the influence of the nature of decisions on the relationship between participation and effectiveness of decisions (see Heller *et al.* 1988).

> *Example*
> Walker and Kalinowski (1994) analyse the contract strategy used and the resulting relationship which exists between the contributors to the Hong Kong Convention and Exhibition Centre.
>   Gibb (2001) investigates applications of standardization and preassembly on construction projects through case study research.
>   Walker and Shen (2002) investigate planning flexibility using a framework of project team understanding and knowledge transfer in two complex projects in Australia.

Pettigrew (1997) suggests that there are '... four forms of case study outputs: the case as an analytic chronology, the diagnostic case, interpretative/theoretical cases, and meta level analysis of writing across a number of case studies'.

Important aspects of case study research are as follows:

(1) Theoretical underpinning must be present – this is applicable to all forms of case study, including descriptive ones.
(2) Case study is a *method* and attention must be paid to associated methodological concepts and procedures.
(3) Case study data can be quantitative and/or qualitative.
(4) Definition of the 'case' and the unit of analysis must be clear. For instance, a study of a contractor's organisation (a single case study) might include a survey of site operatives on different projects within the organisation (embedded units of analysis) and the use of quantitative techniques to analyse the project site data. As long as the major study questions remain at the organisational level, the single organisation remains the major unit of analysis.
(5) In the scientific approach, case study design focuses on empirical testing and covers issues of choice of case(s) (e.g. single-case or multiple-cases; explanatory, descriptive or exploratory case study; case selection) as well as data collection and analysis. Data collection techniques such as interviews and participant observation are not implied.

(6) To enhance validity and reliability of findings under the scientific approach, the case study design has to focus on (a) development of hypotheses based on *a priori* (rival) theories, (b) multiple sources of evidence relying on multiple measures and instruments for empirical testing, (c) development of a case study database.

(7) In testing a theory, case study can be used in conjunction with surveys and other (quasi-) experimental design as part of the methodology.

There is debate concerning how many case studies are necessary to yield a robust piece of research. The answer to the question depends on the purpose of the research and the nature of the case studies undertaken (as noted above). Individual case studies may constitute seminal pieces of work, most obviously if executed in respect of very particular instances (see, e.g. Dyer & Wilkins 1991). In other circumstances, Eisenhardt (1989, p. 545) asserts that generating theory is difficult if fewer than four case studies are used and '. . . empirical grounding [is] likely to be unconvincing unless the case has several mini-cases within . . .' and that while '. . . there is no ideal number of cases, a number between four and ten usually works well'.

Modelling

Classification of models

Mihram (1972) discusses some alternative classification of models – those of Rosenblueth and Weiner (1945), Churchman *et al.* (1957) and Sayre and Crosson (1963).

Rosenblueth and Weiner (1945) categorise models in science as:

- material models: transformations of original physical objects,
- formal models: logical, symbolic assertions of situations, the assertions' representing the structural properties of the original, factual system.

Subcategories, which alternatively may be regarded as alternative categories, are:

- open-box models: predictive models for which, given all inputs, the outputs may be determined,
- closed-box models: investigative models, designed to develop understanding of the actual system's output under different input conditions.

Churchman *et al.* (1957) suggest that models fit into the following categories:

- iconic: visual or pictorial representation of certain aspects of a real system, such as computer screen icons to denote programmes; detail drawings of parts of a building,
- analogue: employs one set of properties to represent some other set of properties which the system possesses (e.g. electrical circuit to mimic heat flow through a cavity wall),
- symbolic: requires logical or mathematical operations (e.g. equation of an 'S curve' of project cash flow).

Sayre and Crosson (1963) suggest the categories of:

- replications: display significant physical similarity to the reality, such as a doll,
- formalisations: symbolic models in which more of the physical characteristics of the reality are reproduced in the model; symbols are manipulated by techniques of a well founded discipline such as mathematics (e.g. $y = a+bx$ is the equation of a straight line),
- simulations: a formalisation model but without entire manipulation of the model by the discipline's techniques in order to yield an analytic solution or a numerical value (e.g. construction project bidding models).

Distillation of the various classifications of models suggests that the common forms of models are:

- iconic
- replication
- analogue
- symbolic.

For research purposes, the more common forms of model are analogue and symbolic, whilst in the construction industry iconic models and replications are usual.

Deterministic and stochastic models

All models contain parameters (variables) which must be identified and quantified for use in the model, together with their inter-relationships. The resultant models are either deterministic – what happened in the past will be replicated in the future – or stochastic (probabilistic) – the laws of probability which governed past realisations will continue to apply in the future. Deterministic models tend to be simpler in form and in the manipulations required than their stochastic counterparts. Whilst, by definition, stochastic models cannot take account of 'shocks' which may occur in reality to the system under study, they are likely to be more realistic but more complex representations. Shocks produce discontinuities ('step changes') in the operation of models – such as changing the 'levels' of inflation forecast by economic models when the oil crises of 1973/4 and 1979 occurred – other shocks may have enduring effects on the operation of the variables in the system being modelled and so, alter the 'pattern' of the model as well as the discontinuity.

Deterministic models assume away random effects, whilst stochastic models seek to incorporate them. A stochastic model will either 'simulate random variables for a whole range of statistical distributions' (Morgan 1984, p. 5) or simulate the particular distribution, if known. For a discussion on the assumptions about distributions and their use in construction project bidding models and risk management see Chau (1995, 1997) and Fellows (1996). Whether deterministic or stochastic, a model should mimic the effects present in the reality by inclusion in the constituents of the model. For deterministic models, this may be an 'express' residual element which is not included in any other component. For stochastic models it may mimic reality by incorporation of probabilities, whether or not a distribution type is assumed or determined.

For realities in which changes occur only slowly or consistently, deterministic models can be appropriate – the pattern of any consistent change can be determined and incorporated to yield dynamism to an otherwise static model, such as in the deterministic analysis of time series discussed in Chapter 7. Similarly, dynamism can be incorporated in stochastic models. Dynamic components of models may be continuous, as in a growth trend, or discrete, as in seasonal elements of construction workload. 'Most systems ... develop their characteristic behaviour over time, so that any ... models of this behaviour needs to be ... dynamic ...' (Mihram 1972).

The modelling process

Models may be used to investigate and/or to predict; for managers, predictive models are more valuable, whilst auditing requires investigative modelling. PERT is a stochastic predictive model; an investigative model could comprise a set of equations in several unknowns, sufficient that provided a certain number of values of some of the variables are known, the equations can be

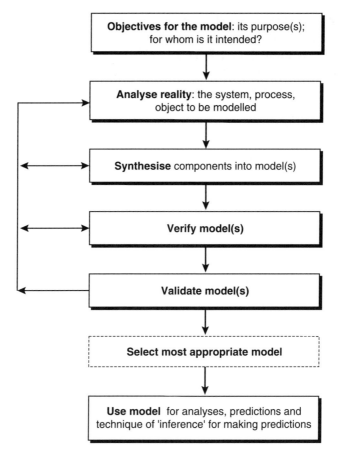

Fig. 4.2   The modelling process (developed from Mihram 1972).

used to determine the remainder, as in linear programming. The modelling process is depicted in Fig. 4.2.

Verification of a model involves determining whether the structure of the model is correct; this is achieved by testing the model through examining the outputs resulting from the model under a given set of inputs. The model is verified if the outputs are appropriate, i.e. they approximate to 'expectations' of what a good model of the reality would yield. Models which are verified may pass to the next stage, *validation*, whilst those which are not verified may be discarded or, especially if only one model is being examined, be returned to the analysis or synthesis stages for further scrutiny and amendment. In any testing, it is essential that data are used which have not been employed in building the model. Thus, the tests are independent and valid. (If data employed to build the model are used for testing, there is an element of a 'self-fulfilling prophecy', thereby rendering the testing invalid.)

In validating a model, the model's output resulting from known inputs is compared to realisations of the reality, such as an ex-post forecast (see Fig. 5.1, p. 132). If possible, it is helpful to carry out such validatory testing for several sets of inputs and known outputs of the reality to examine consistency of the model over a range of conditions, preferably including 'extremes'. If a number of models have been suitably verified, it is usual for validation to be used to select the most appropriate models. Verification may suggest a model which is 'best' on the basis of theoretical 'fitting' – from criteria based on analysis of the model, whilst validation may yield a different 'best' model on the basis of closest fit of output to test realisations. The choice of model will depend on the objective of the modelling exercise, its use and by whom it will be used etc., and the differences in the two forms of performance between the models being tested. In the final stage, the verified and validated model may be used. However, the model also may be iterated so that inferences can be considered about the possibility of extending the operation of the model to other conditions, by relaxing some of the restrictions or assumptions for the models' appropriate operation and use.

Simulation

Dynamism

Simulation involves the use of a model to represent the essential characteristics of a reality, either a system or a process. So, whilst a model may be a static representation, such as an architectural model, a simulation involves some element of dynamism, if only because it models a process rather than an object. Thus, flight simulators mimic the behaviours of aircraft under specified conditions; models in wind tunnels are used to investigate the flow of air (simulations with scaling) about the model. Hence, often, simulation is used to examine how the behaviour of the reality is likely to change consequent upon a change in the values of input variables etc. in the representative model.

Simulation is used to assist prediction of the behaviour of a reality or/and to revise a model to enhance its predictive accuracy or predictive capability. Morgan (1984) suggests a variety of purposes for simulation:

- explicitly mimic the behaviour of a model
- examine the performance of alternative techniques
- check complex mathematical/analytic models

- evaluate the behaviour of complex random variables, the precise distribution(s) of which is (are) unknown.

Heuristics

Clearly, as modelling lies at the base of simulation, formally in many instances but informally in others, the procedures employed to obtain an appropriate model of a reality apply to simulations. Increasingly, computers are used in simulations; often, such simulations use *heuristics*, or 'rules of thumb' in their replicating behaviour of a reality as many realities comprise a complex system of independent but interactive components which act together. Neither the individual components and their behaviours, nor the interactions of those components may be known, understood and modelled in detail or exactly, so, if only due to practical constraints such as time, cost etc., a sufficiently accurate simulator may be produced through observation and measurement of the reality. Heuristics for the simulator may be produced through interpretation and interpolation, deduction and induction, using knowledge of appropriate theory and principles.

Mitchell (1969) notes that, 'in a practical context ... [simulation can] ... be an extremely powerful way of understanding and modelling a system. More importantly, from the practical point of view, it can be the only way of deriving an adequate model of a system'. Meier *et al.* (1969) extend the assertion, stating that 'experimentation', by means of computer simulation, can overcome some of the restrictions that exist when other forms of analysis are used. By using representative heuristics, simulation, 'opens up the possibility of dealing with the dynamics of the process, too complex to be represented by more rigid mathematical models.... Simulation may make possible experiments to validate theoretical predictions of behaviour in cases where experimentation on the system under study would be impossible, prohibitively expensive or complicated by the effect of interaction of the observer with the system ...'.

Approaches

Whilst simulation may be used to represent the behaviour of a precise model in a realistic way, because it is a model in which the natures or distributions of the dynamic elements are known, as for parametric statistics, simulation also may be used to evaluate the behaviour of a system of random variables of which the distributions are unknown, as for non-parametric statistics. In the latter case, theoretical, experimental and experiential evidence may be used to suggest appropriate distributions to be used in simulations, although considerable debate over the appropriate distributions may ensue (see Chau 1995, 1997 and Fellows 1996).

> *Example*
> Frequently, the Monte Carlo technique of random number selection is used in simulations. About 1000 iterations is normal in computer simulations employing Monte Carlo for predicting durations, costs and prices of construction activities.
>   Whilst the form of the assumed distribution of the random variable(s) may influence the result of the simulation, it is acknowledged that the limiting values used for the range of possible values which the variables may take is important. Such a view is significant in PERT's use of the β-distribution and the determination of each activity's limiting values, optimistic and pessimistic, which may be determined by many techniques – Delphi, analyses of past performance, expert estimation etc.

Game approaches are popular as simulations for people making decisions. Business games are used to enable players to see the consequences of decisions made in a simulation of an organisation operating in an environment, using heuristics as its operating rules – for both environmental and organisational changes and outcomes. Clearly, learning by the players is a major objective, but games may be used in similar ways to replicate organisations and to make predictions – such as for alternative 'futures'.

> *Example*
> A major gaming simulator for operating a construction company is AROUSAL (developed by Peter Lansley, University of Reading, UK) which provides iterations of bidding, resourcing projects and running the company over a pre-determined period of the company's operation; projects available to bid are offered, results provided, accounts for the projects and company produced, personnel profiles for employment and of performances are available periodically etc. Such gaming allows managers to practice and develop skills in a realistic but 'safe' context.

Generally, games have expressed or assumed objectives and, sometimes, both. They have rules, but the outcomes are determined by inputs and by circumstances, which are random variables. Hence, the need to appreciate risks, at least as outcome variabilities are demonstrated, plus how and to what extent they may be influenced. Frequently, *Partnering* is cited as a win-win (or non-zero-sum) game because, if the participants, partners or players adhere to the rules of behaviour, all are asserted to benefit. Negotiating a final account under 'traditional' project arrangements is a win-lose (zero-sum) game, since what one party gains, the other must lose and because the aggregate is a constant, it is the pattern of sharing that aggregate which is at stake.

Simulation offers a unique opportunity to observe the dynamic behaviour of complex, interactive systems. A carefully constructed, realistic simulation model provides a laboratory environment in which to make observations under controlled conditions for testing hypotheses, decision rules and alternate systems of operation under a variety of assumed circumstances.

Summary

It is important that empirical work should not begin until the review of theory and literature has been carried out. Problems of placing too much reliance on experience have been considered. Issues in research design are examined. The main approaches to collecting data for qualitative studies have been reviewed, notably the use of 'grounded' theory. Similarities between making measurements and collecting quantitative data were noted; problems of data allocation and coding were reviewed. Requirements in modelling and difficulties in conducting empirical study in a social context have been considered – especially for the handling of different types of variables and establishing relationships between them, in particular, the issue of causality.

References

Babbie, E. (1992) *The Practice of Social Research*, Wadsworth, Belmont, CA.

Back, L. (1998) Reading and writing research. In: *Researching Society and Culture* (ed. C. Seale), pp. 285–296, Sage Publications, London.

Baron, R.M. & Kenny, D.A. (1986) The moderator–mediator variable distinction in social psychological research: conceptual, strategic, and statistical considerations, *Journal of Personality and Social Psychology*, **51**(6), 1173–1182.

Bogdan, R.C. & Biklen, S.K. (1982) *Qualitative Research for Education*, Allyn and Bacon, Boston.

Charmaz, K. (1983) The grounded theory method: an explication and interpretation. In: *Contemporary Field Research* (ed. R.M. Emerson), Little, Brown, Boston.

Chau, K.W. (1995) Monte Carlo simulation of construction costs using subjective data, *Construction Management and Economics*, **13**(5), 369–383.

Chau, K.W. (1997) Monte Carlo simulation of construction costs using subjective data: response, short communication. *Construction Management and Economics*, **15**(1), 109–115.

Churchman, C.W., Ackoff, R.L. & Arnoff, E.L. (1957) *Introduction to Operations Research*, Wiley, New York.

Clegg, S.R. (1992) Contracts cause conflicts. In: *Construction Conflict Management and Resolution* (eds P. Fenn & R. Gameson), E. & F.N. Spon, London.

Condray, D.S. (1986) Quasi-experimental analysis: a mixture of methods and judgement. In: *Advances in Quasi-experimental Design and Analysis* (ed. W.M.K. Trochin), Jossey-Bass, San Francisco.

Dyer, W.G. & Wilkins, A.L. (1991) Better stories, not better constructs, to generate better theory: a rejoinder to Eisenhardt, *Academy of Management Review*, **16**(3), 613–619.

Easterby-Smith, M., Thorpe, R. & Lowe, A. (1994) *Management Research: An Introduction*, Sage Publications, London.

Eisenhardt, K.M. (1989) Building theories from case study research, *Academy of Management Review*, **14**(4), 532–550.

Fellows, R.F. (1996) Monte Carlo simulation of construction costs using subjective data: a note, *Construction Management and Economics*, **14**(5), 457–460.

Fine, B. (1975) Tendering strategy. In: *Aspects of the Economics of Construction* (ed. D.A. Turin), George Godwin, London.

Geertz, C. (1973) *The Interpretation of Cultures*, Fontana, London.

Gibb, A.G.F. (2001) Standardization and pre-assembly – distinguishing myth from reality using case study research, *Construction Management and Economics*, **19**(3), 307–315.

Glaser, B.G. & Strauss, A.L. (1967) *The Discovery of Grounded Theory: Strategies for Qualitative Research*, Aldine, Chicago.

Gould, S.J. (1988) *The Mismeasure of Man*, Penguin Books, London.

Hammersley, M. & Atkinson, P. (1983) *Ethnography: Principles and Practice*, Tavistock, London.

Harris, L.C. & Ogbonna, E. (2002) The unintended consequences of cultural interventions: a study of unexpected outcomes, *British Journal of Management*, **13**(1), 31–49.

Heller, F.A., Drenth, P.J.D., Koopman, P.L. & Rus, V. (1988) *Decisions in Organizations: A Three Country Comparative Study*, Sage, London.

Hicks, C.R. (1982) *Fundamental Concepts in the design of Experiments*, 3rd edn, Holt-Saunders International, Philadelphia.

Hughes, C. (1994) From field notes to dissertation: analysing the stepfamily. In: *Analysing Qualitative Data* (eds A. Bryman & R.G. Burgess), pp. 35–46, Routledge, London.

Leary, M.R. (1991) *Behavioural Research Methods*, Wadsworth, Belmont, CA.

Leary, M.R. (2004) *Introduction to Behavioural Research Methods*, 4th edn, Pearson Education Inc., Allyn and Bacon, MA.

Levin, R.I. & Rubin, D.S. (1991) *Statistics for Management*, 5th edn, Prentice-Hall, Englewood Cliffs, NJ.

Luthans, F. (1992) *Organizational Behaviour*, McGraw-Hill, New York.

Meier, R.C., Newell, W.T. & Pazer, H.L. (1969) *Simulation in Business and Economics*, Prentice-Hall, Englewood Cliffs, NJ.

Mihram, G.A. (1972) *Simulation: Statistical Foundations and Methodology*, Academic Press, New York.

Mitchell, G.H. (1969) Simulation, *Bulletin of the Institute of Mathematical Applications*, **5**(3), 59–62.

Morgan, B.J.T. (1984) *Elements of Simulation*, Chapman & Hall, London.

Morse, J.M. (1994) Emerging from the data the cognitive process of analysis in qualitative enquiry. In: *Critical Issues in Qualitative Research* (ed. J.M. Morse), Sage, London.

Oakley, J. (1994) Thinking through fieldwork. In: *Analysing Qualitative Data* (eds A. Bryman & R.G. Burgess), Routledge, London.

Petersen, R.G. (1985) *Design and Analysis of Experiments*, Marcel Dekker Inc., New York.

Pettigrew, A.M. (1997) What is processual analysis?, *Scandinavian Journal of Management*, **13**(4), 337–348.

Reber, A.S. (1995) *Dictionary of Psychology*, 2nd edn, Penguin, Harmondsworth.

Remenyi, D., Williams, B., Money, A. & Swartz, E. (1998) *Doing Research in Business and Management. An Introduction to Process and Method*, Sage, London.

Rooke, J., Seymour, D.E. & Fellows, R.F. (2003) The claims culture: a taxonomy of attitudes in the industry, *Construction Management and Economics*, **21**(2), 167–174.

Rooke, J., Seymour, D.E. & Fellows, R.F. (2004) Planning for claims: and ethnography of industry culture, *Construction Management and Economics*, **22**(6), 655–662.

Rosaldo, R. (1989) *Culture and Truth: the Remaking of Social Analysis*, Routledge, London.

Rosenblueth, A. & Weiner, N. (1945) The role of models in science, *Philosophy of Science*, **12**, 316–321.

Runkel, P.J. & McGrath, J.E. (1972) *Research on Human Behavior*, Holt, Rinehart and Winston, New York.

Sayre, K.M. & Crosson, F.J. (eds) (1963) *The Modelling of Mind: Computers and Intelligence*, Simon & Schuster, New York.

Schatzman, L. & Strauss, A.L. (1973) *Field Research: Studies for a Natural Sociology*, Prentice-Hall, Englewood Cliffs, NJ.

Seale, C. (1998) *Researching Society and Culture*, Sage Publications, London.

Silverman, D. (1998) Analysing conversation. In: *Researching Society and Culture* (ed. C. Seale), pp. 261–274, Sage Publications, London.

Slater, D. (1998) Analysing cultural objects: content analysis and semiotics. In: *Researching Society and Culture* (ed. C. Seale), pp. 233–244, Sage Publications, London.

Strauss, A.L. (1987) *Qualitative Analysis for Social Scientists*, Cambridge University Press, Cambridge.

Strauss, A.L. & Corbin, J. (1990) *Basics of Qualitative Research: Grounded Theory Procedures and Techniques*, Sage, Newbury Park, CA.

Tesch, R. (1991) Software for qualitative researchers, analysis needs and program capabilities. In: *Using Computers in Qualitative Research* (eds N.G. Fielding & R.M. Lee), Sage, London.

Tonkiss, F. (1998) Analysing discourse. In: *Researching Society and Culture* (ed. C. Seale), pp. 245–260, Sage Publications, London.

Turner, B.A. (1994) Patterns of behaviour: a qualitative enquiry. In: *Analysing Qualitative Data* (eds A. Bryman & R.G. Burgess), pp. 195–215, Routledge, London.

Van Maanen, J. (1979) The fact and fiction in organizational ethnography, *Administrative Science Quarterly*, **24**(4), 539–550.

Walker, A. & Kalinowski, M. (1994) An anatomy of a Hong Kong project organization, environment and leadership, *Construction Management and Economics*, **12**(3), 191–202.

Walker, D.H.T. & Shen, Y.J. (2002) Project understanding, planning, flexibility of management action and construction time performance: two Australian case studies. *Construction Management and Economics*, **20**(1), 31–44.

Walsh, D. (1998) Doing ethnography. In: *Researching Society and Culture* (ed. C. Seale), pp. 217–232, Sage Publications, London.

Wikipedia (2007) http://en.wikipedia.org/wiki/Intervening_variable, accessed 21 June 2007.

Yin, R.K. (1993) *Applications of Case Study Research*, Sage, Newbury Park, CA.

Chapter 5

Hypotheses

The objectives of this chapter are to:

- discuss the roles of hypotheses;
- examine the essentials of a valid hypothesis;
- emphasise the necessity for *objective testing of hypotheses*;
- demonstrate the *role of sampling* for testing hypotheses;
- examine common statistical measures and the testing of null hypotheses;
- discuss *validities*.

Essentials of a valid hypothesis

Not all research projects require a hypothesis.

It is appropriate to include a hypothesis in research when it is based on theory and previous work and, being so, sets out to test the existence of certain variables and/or any relationship between them. Quantitative studies are the most obvious instances of research projects which have a hypothesis to test.

For a great deal of research, notably, the application of scientific method, or quantitative studies, it is both possible and important to draw on theory and literature (findings of executed research) to formulate hypotheses to be tested. In both producing a proposal for a research project and in executing the research, the hypothesis acts as the focus for the work and one which helps to identify the boundaries also.

In management, engineering and other applied disciplines, it is usual for a hypothesis to be appropriate and very useful in delineating and guiding the study. If it is appropriate to develop a hypothesis for testing via the research, it is a good idea to do so.

The *Oxford English Dictionary* (OED 2007) defines a hypothesis in a number of (similar) ways:

- A supposition in general; something supposed or assumed to be true without proof or conclusive evidence; an assumption.
- A supposition or conjecture put forth to account for known facts; *esp.* in the sciences, a provisional supposition from which to draw conclusions that shall be in accordance with known facts, and which serves as a starting-point for further investigation by which it may be proved or disproved and the true theory arrived at.
- A proposition or principle put forth or stated (without any reference to its correspondence with fact) merely as a basis for reasoning or argument, or as a premiss from which to draw a conclusion; a supposition.

A proposition is (OED 2007):

- Something proposed for discussion or solution; a problem, a riddle; a parable.
- A statement which is capable of truth or falsity; (also occas.) a mental formulation of that which is expressed by such a statement.

'... propositions state the relations among constructs, and on the more concrete level, hypotheses (derived from the propositions) specify the relations among variables' (Bacharach 1989). Thus, 'The primary difference between propositions and hypotheses is that propositions involve concepts, whereas hypotheses require measures'.   (Whetten 1989)

A theory is a much more definite statement than a hypothesis or a proposition; due to the results of testing, a theory has gained general acceptance as being an accurate explanation of a process (at least, the best explanation found to date). A theory is (OED 2007):

- A conception or mental scheme of something to be done, or of the method of doing it; a systematic statement of rules or principles to be followed.
- A scheme or system of ideas or statements held as an explanation or account of a group of facts or phenomena; a hypothesis that has been confirmed or established by observation or experiment, and is propounded or accepted as accounting for the known facts; a statement of what are held to be the general laws, principles, or causes of something known or observed.

- That department of an art or technical subject which consists in the knowledge or statement of the facts on which it depends, or of its principles or methods, as distinguished from the *practice* of it.
- In loose or general sense: A hypothesis proposed as an explanation; hence, a mere hypothesis, speculation, conjecture; an idea or set of ideas about something; an individual view or notion.

Although the last definition is in common use, it is not helpful in the context of research!

A hypothesis is *a statement*, a conjecture, a hunch, a speculation, an educated guess. Usually, however, a hypothesis is a speculative statement which is a reasonable suggestion of a (causal) relationship between the independent variable and the dependent variable. In cases of more complex anticipated relationships, the hypothesis may include the anticipated impact of other variables (mediators, moderators, intervening) on the main relationship. The basis for a hypothesis may be theory, practice, experience or some combination of those and so, the hypothesis may be formulated during the very early stages of producing a proposal (if based on experience) but, in many cases, it will be formulated only after preliminary investigation of theory and literature.

In framing (phrasing) a hypothesis, it is useful to bear in mind the formal notion of a 'null' hypothesis and its complement, the 'alternate'(or alternative) hypothesis; together, the null and alternate hypotheses cover all possibilities of the relationships between the variables involved. Thus, in an experiment involving tossing a coin to determine whether the coin is fair (i.e. unbiased – an equal chance of 'heads' and of 'tails' for each toss):

'null' hypothesis – H_O = the coin is fair
'alternate' hypothesis – H_A = the coin is not fair (i.e. the coin is biased).

The KISS principle is an efficiency principle – Keep It Simple, Stupid. It is a good research principle from several views – keep contents straightforward, simple and clear; write plainly so that the language is easy to read and the arguments can be followed readily and have no ambiguities and, perhaps most fundamentally, search for the simplest basic facts, principles, models etc. This is what mathematicians call *elegance* (a hypo-thesis, *not* a 'hippo-thesis').

The overall aim of the research (or the research question(s)/problem) may be articulated further as a proposition which, for research operation, may be expressed in greater detail as a (small) set of component hypotheses, each of which relates to a research objective.

Sometimes, it will be necessary to supplement a main hypothesis with sub-hypotheses which are components of the main hypothesis; that occurs where complex phenomena are being investigated such that the overall (strategic) relationship cannot be investigated directly but must be addressed through components or observable phenomena (such as investigating culture through examining behaviour, language and other manifestations of the underlying culture construct). That raises the need for awareness of the issues regarding reductionism and holism. In other instances, a hypothesis may be supplemented with *auxiliary statements*.

From a research perspective, to be valid and useful, a hypothesis must comply with three primary requirements:

- It must be *testable* – so that it may be supported or rejected from empirical evidence.
- It must be *positive* – testing the hypothesis concerns what is, *not* what ought to be (*normative*; as normative statements are value judgements).
- It must be expressed in *clear and simple language* – so that it means the same to everyone (i.e. it is consistent; *of constant meaning*).

Indeed, Popper (1989) has argued that a hypothesis must be *falsifiable* if it is to qualify as a hypothesis for scientific research. The results of testing a hypothesis may support the hypothesis, disprove/disconfirm/not support the hypothesis, or be inconclusive. If a hypothesis is supported, wherever possible (notably, if support is found via statistical testing) the level of confidence of the support should be stated. Popper also notes that a theory or a proposition must admit the possibility of being shown to be false if it is to be classified as scientific.

A hypothesis which is confirmed adds support to the body of knowledge and theory and may, if supported by related findings over time, develop into a theory itself (as in the development of knowledge discussed in Chapter 1 – the dialectic triad of thesis, antithesis, synthesis).

Whenever it is decided that it is appropriate to include a hypothesis in a research project, it is likely that a variety of hypotheses will be considered. Apart from ensuring that the (main) hypothesis adopted relates to the focal variables of the research, the selection of which hypothesis to employ should be guided by the following considerations (Schick & Vaughn 2002):

- Testability – that the hypothesis can be falsified through appropriate testing (data collection and analysis).
- Simplicity – that the hypothesis is parsimonious, follows the KISS principle, adheres to 'Occam's Razor' (minimise assumptions – remove those which make no material difference to the outcome).
- Scope – the possible application of the hypothesis to an array of situations.

- Fruitfulness – the potential for application of the hypothesis to other phenomena in the future.
- Conservatism – compatibility of the hypothesis with existing theory and literature.

Roles of hypotheses

In conjunction with the objectives, a hypothesis specifies the focus of the research. However, not all research projects will have hypotheses to be tested. Research which is exploratory, especially in new contexts, may have the formulation of a hypothesis as one of its main objectives for the outcomes of that research. Thus, a hypothesis is inappropriate for a qualitative study which seeks to carry out a fundamental investigation to identify what is occurring – such as to observe behaviour in a highly novel environment or in a new community in which established values, theories etc. may not apply.

For research which is a fundamental exploration of human behaviour in a topic for which theory has not been developed to any significant extent and, hence, where little or no research has been carried out, it is not possible to generate hypotheses in a meaningful way. The appropriate approach in such circumstances is not to force issues artificially, in this case identification of variables and supposing causal relationships between them, but to observe behaviour whilst minimising the effects of such observation on the subjects' behaviours.

Moser and Kalton (1971) note – 'Surveys thus have their usefulness both in leading to the formulation of hypotheses and, at a more advanced stage, in putting them to the test. Their function in a given research project depends on how much is already known about the subject and the purpose for which the information is required.' They warn against the forced inclusion of hypotheses in surveys of 'factual enquiries' which should be undertaken only after thought has been given to what should be asked. Forcing hypotheses into such studies inappropriately results in only trivial hypotheses and, thereby, hardly avoids the (poorly informed) criticism of 'factual surveys' that they do not include hypotheses and their testing! So, it is important to recognise when it is appropriate to include hypotheses in the research and when it is not.

As the early stages of research progress, from the preliminary investigations undertaken to help to produce the proposal, with the review of theory and literature, the main hypothesis and sub-hypotheses may be modified as greater knowledge of the topic and main issues involved is gained. On occasions, especially when working in new areas of study, some researchers prefer not to include a hypothesis in the proposal but to develop the hypotheses once the theory and literature have been examined, such that the hypothesis arises out of a thorough and up-to-date understanding of the topic.

Provided there is sound appreciation of the dynamism of any research project, which is due to the very nature of the research process, the suggestion of an initial hypothesis, which may be subject to revision, has certain desirable elements. It is essential to recognise that as knowledge advances and increases, so beliefs change, and to adhere to initial ideas rigidly excludes advances. Thus, within a paradigm, knowledge advances produce 'drift' within that knowledge framework until a point is reached where a 'step-change' paradigm shift occurs, analogous to the dialectic triad of thesis–antithesis–synthesis (see Chapter 1).

For research to yield its full potential, the developmental aspects must be incorporated. It is important that researchers do not feel either forced to include a hypothesis in a proposal or to be restricted by not being allowed to amend a hypothesis once it has been advanced. However, if too much flexibility is allowed over the provision and/or adaptation of hypotheses, the objectives of the work, as manifested in outputs, may suffer, and it may be that the research hypotheses change constantly from the original, with the result that none are ever tested.

Hence, there is a consideration of what is appropriate – not too much forcing but also a degree of limitation, so that whatever approach is adopted as being suitable for the research project, that approach is followed with rigour. What is done, how, when etc. must be capable of being substantiated by the researcher, who should be able to provide valid reasons for the approach – including changes made to the research plan. Do not do things without good reason. Reasons of practicality can be just as valid and important as reasons of theory.

Objective testing of hypotheses

A hypothesis is the main statement of supposition permissible which will be tested rigorously by the research to remove as much of the supposition or uncertainty as possible and replace it with knowledge, this may be certainty – more realistically, risk or probability. Many research projects are dependent upon assumptions, often due to the nature of the theory on which the research is based. Such assumptions are suppositions and, as such, must be identified and stated expressly in the research. If possible, investigating their validity should be conducted and, perhaps more especially, the consequences of relaxing those assumptions on the results of the research.

Chambers English Dictionary defines a hypothesis as:

- a supposition
- a proposition assumed for the sake of argument
- a theory to be proved or disproved by reference to facts
- a provisional explanation of anything.

The fourth definition encapsulates the view of the highly qualitative, ethno-graphic type of research which seeks to observe in order to establish hypotheses; provisional explanations which will be subjected to testing for verification.

The third definition of a hypothesis raises a significant philosophical prob-lem (this is not an issue of semantics). Commonly, especially new, researchers believe they should prove or disprove a hypothesis and so they set out to do so. Proving, in the absolute sense, will not be possible, nor should that be a goal as such a goal is likely to introduce *bias*. The goal is to *test* the hypoth-esis as rigorously as possible; that means *objectivity*. In testing a hypothesis, a researcher should seek to provide evidence, through results of the testing, to support or reject the hypothesis at an appropriate level of confidence and obtained, if possible, through statistical testing and, therefore, quantified. Any level of confidence calculated should be specified in the results – that aids appreciation of the conclusions and their validity.

For statistical testing, of null hypotheses, the phrasing of the null hypoth-esis indicates whether a 'one-tail test' (the value is either greater than, or is less than, some specified, critical value to distinguish the null and alternate hypotheses) or a 'two-tail test' (different from the content of the null hypothe-sis – the alternate hypothesis is supported of test results, which show that the value is either greater or less than the value specified in the null hypothesis).

In examining the results of statistical testing, it is useful to consider the following sequence of questions:

- Are the results statistically significant?
- What is the level of significance (confidence level)?
- What are the numerical results (e.g. coefficients of correlation)?
- What do the results mean (with reference to theory and literature to identify causality)?

Role of sampling

Usually, testing a hypothesis involves sampling from the population by collecting data and executing analyses. Hence, statistical techniques play a

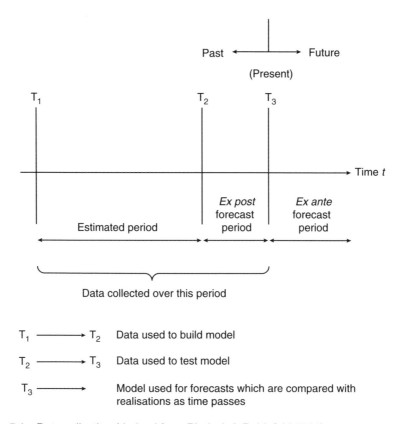

Fig. 5.1   Data collection (derived from Pindyck & Rubinfeld 1981).

very important role in testing hypotheses. There are many statistical techniques which can be employed to test hypotheses. Although several of the more common statistical techniques are considered in this book, specialist statistical texts should be studied to gain appreciation of the range of techniques available and to ensure that the most appropriate are used. Further guidance may be obtained from the manuals which accompany computer software statistical packages. Fig. 5.1 emphasises appropriate data sets to be used for building, testing and, potentially, modifying 'models' and the relationships between variables.

As time passes and more data become available, those data may be employed to modify the initial model continuously. However, it is important that data used to build a model are *not* also used to test the model; the testing must be done against data not used in the model-building, otherwise the model would, inevitably, contain an element of being a 'self-fulfilling prophesy' with consequent distortion to the fit, variability and accuracy measurements.

Further, it should be noted that many measurements are really indicators of varying degrees of error (from the 'true' measure). In dynamic analyses, such as for an economy, indicators are classified as:

- *Leading* (long; short) – 'where we think we will be'.
- *Co-incident* – 'where we think we are'.
- *Lagging* (short; long) – 'where we think we have been'.

Sampling is necessary because it is rarely possible to examine an entire population. It would be possible to survey the techniques used to pre-stress pre-cast concrete components manufactured in Hong Kong as the population of manufacturers of such components is small. It would *not* be possible to carry out a full population survey of the type of tea preferred by the Hong Kong population. The main reasons for the impossibility include population changes (births, deaths, migrations), time, cost, tracing people, obtaining the data. Fortunately, to obtain a good representative picture of the population's tea-drinking preferences, it would be possible to use a sample of the population, which is much smaller than the total population but sized and structured to be statistically representative. Clearly, the results obtained from such sampling will not be exactly the same as if the whole population had been consulted, but the result will be adequate for the purpose for which the information is required, such as for an importer to know the appropriate quantities of each type of tea to import annually into Hong Kong.

So, the sample needs to be representative of the population in order to produce a result of theoretical and practical value and that the results obtained from the sample approximate as closely as possible to those which would be obtained if it were possible to survey the entire population. Thus, sampling is designed so that the sample is likely to be sufficiently representative of the population in order that the results obtained from the tests, via statistical inference, have sufficient external validity to be applied to the population at a given and stated level of confidence. (See also Fig. 4.2, p. 117.)

Common statistical measures

Statistics may be classified in several ways. One important, basic classification is 'descriptive statistics' and 'analytic statistics'. Descriptive statistics are used to express the important features of a population (population *parameters*), intended sample or sample obtained (sample *statistics*) numerically; they include percentages (e.g. response rates) and other numerical descriptors of the distribution (e.g. of the sample of responses) under examination – including measures of centrality (means, mode, median) and spread/variability (variance, standard deviation). Use of descriptive statistics does NOT constitute analysis!

A very common measurement is the average; in statistics, there are two important measurements of the average or *mean*:

- the *arithmetic mean*; which most people understand as the average
- the geometric mean.

The arithmetic mean is expressed mathematically as:

$$\bar{x} = \frac{1}{n}\sum_{i=1}^{n} x_i \quad \text{or} \quad \bar{x} = \frac{\sum x}{n}$$

where: n = no of items
x_i = an item of value i where the value of i ranges from 1 to n
$\bar{x}$ = arithmetic mean.

The geometric mean is calculated as:

$$GM = \sqrt[n]{x_1 x_2 \ldots x_n}$$

In sampling, the mean of the sample is derived to approximate, as closely as practicable, to the mean of the population. Distributions are very important too – averages can be very misleading. ('Geometric' growth pattern applies for quantities which change over time (e.g. rate of inflation) and gives the average quantity for a period under review.)

> *Example*
> A statistician once illustrated the point: 'if I have one foot in a bucket of ice and the other foot in a bucket of boiling water, on average, I'm OK!'
>
> Is s/he?

Clearly the statistician would not be 'OK', so it is important to consider measurements of dispersion, distribution or variability too. The *median* is the mid point of a distribution. As extreme or 'outlier' values may distort the mean, the median may be a preferable indicator of such a distribution's central tendency. The *mode* is the value which occurs most frequently in a distribution. A distribution may have more than one mode.

Example
(a) Symmetrical distribution (e.g. normal distribution)
     Mean, mode and median coincide.

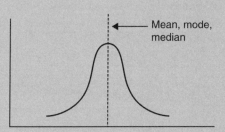

(b) Skewed distribution
     Mean and, to a lesser extent, median move away from the mode in the
     direction of the skew (long tail).

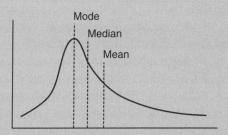

Note: 'Outlier' points of the distribution affect the location of the mean more
than the median and the median more than the mode. Hence, if 'outliers'
cannot be disregarded, their effects must be acknowledged.

(c) Bimodal distribution

Note: The modes need not be of equal size or value.

The arithmetic mean is the value which, numerically, is the most representative of the series, or distribution. The geometric mean is used where measurements of relative changes are being considered. The geometric mean is used in retail price indices; for such variables, it is usual to be interested in the percentage change between successive periods.

Example

The marks gained by a group of students form a frequency distribution (frequencies being the number of students obtaining each mark); for a frequency distribution, the arithmetic mean is:

$$\bar{x} = \frac{\sum fx}{\sum f}$$

where: $\sum fx$ is the sum of the number of students obtaining each mark
         multiplied by the marks obtained
         $\sum f$ is the total number of students.

Samples, and the populations from which samples are drawn and which the samples should represent, each have statistical distributions – commonly assumed to be normal; tests for normality should be applied.

However, many populations are structured in a particular way which is important to the research. In order to obtain a sample which is representative of the population, the sample should reflect the structure of the population closely. The structure of the population yields the sampling frame – the structure used to determine the form of the sample from which the data will be gathered.

Generally, sample surveys have the objective of either:

- estimating a parameter in the population – the estimate made should be accompanied by a note of its precision
- testing a statistical hypothesis concerning the population – the results of testing will lend support to or lead to rejection of the hypothesis; such tests require a criterion against which deviation of the result from the sample can be judged against the value hypothesised.

In both instances, the measure of precision used is the standard error.

It is important to note that statements which are made on the basis of results derived from samples are probability statements and that random sampling should be employed. Random sampling is where each member of a population has a known and non-zero probability of being included in the sample. If a sample of size n is taken an infinite number of times from a population by a random sampling method, the distribution of the sample means is the sampling distribution of the mean.

An *estimator*, the method of estimating the population parameter or measurement from the sampled data, should be unbiased; it is unbiased if, on average over an infinite number of samples, the *sample estimates* (measurements yielded by the estimator, such as the sample mean, $\bar{x}$) equal the *population parameter* (in this case, the population mean, μ).

Systematic error, often called bias, is when the errors assume a regular pattern of under- or over-measurement by a proportion or an amount; such error can be revealed by checking and can be compensated by an adjustment. Further analysis of the errors may reveal source(s) as well as size to such an extent that it may be possible to reduce or even eliminate the error. Systematic error should be avoidable. *Unsystematic error*, or random error, is almost inevitable, but its size should be kept to a practical minimum by research design, rigour of execution and checking.

The *standard error* of the means measures the degree to which estimates which are obtained from different samples are likely to differ from each other. For a finite (quite small) population:

$$\text{Standard error } (\bar{x}) = \sqrt{\frac{\sigma^2}{n} \times \frac{N - n}{N - 1}}$$

where: σ = standard deviation of the population parameter
N = size of the population
n = sample size.

For 'infinite' population sizes, reduces to:

$$\sigma_{\bar{x}} = \frac{\sigma}{\sqrt{n}}$$

As illustrated in Fig. 5.2, the inaccuracy of an estimate obtained from a sample is measured by the extent to which it differs from the population parameter; an estimator is unbiased if the mean of the estimates obtained from all possible samples is equal to that of the population parameter. The standard error is a measure of the fluctuations of estimates around their own mean (shown by the sampling fluctuations). Hence, the reference point of the standard error is the expected value rather than the population value, as it is a 'sample-based' measure.

The *mean square error* (MSE) measures the variability around the population value:

$$\text{MSE} = Var(\bar{x}) + (\mu - m)^2$$

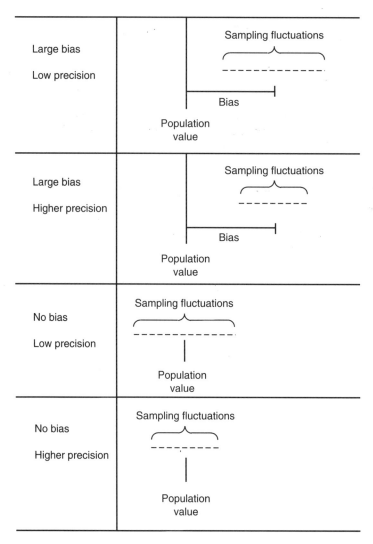

Fig. 5.2   Bias and precision (*source*: Moser & Kalton 1971, following Deming 1950).

where: $Var(\bar{x})$ is the variability around the expected value
μ is the population value
m is expected value of the estimator; the mean of the estimates
obtained from all possible samples
$(\mu - m)$ is the bias of the sample.

MSE is the arithmetic average (mean) of the errors squared (each one is multi-plied by itself) which thereby eliminates the sign of the error terms. The result is a 'magnified' measure of the average error of the measures made. The utility of the MSE of the sample is that it provides an unbiased estimate of the variance of the population (σ^2). To re-scale the measure to the same scale as the sample and population, the square root of MSE is calculated to give the 'root mean square error'.

Normal distribution

Much of statistics uses a particular distribution which has been found to occur very commonly – the *normal distribution*. The normal distribution is a probability density function which has certain, particular properties.

Fig. 5.3 shows the normal distribution. The main features of the normal distribution are:

- measurements of the random variable, X, are on an interval scale or a ratio scale,
- values of X are symmetrical about the mean, μ,
- the distribution has a central tendency; i.e. the closer values of X are to μ, the more frequently they occur,
- the tails of the distribution spread to infinity.

Thus, the distribution is 'bell-shaped'.

Another important measurement for a normal distribution concerns the variability or spread of the distribution. That measurement is the *variance*; the square root of the variance is the standard deviation, σ.

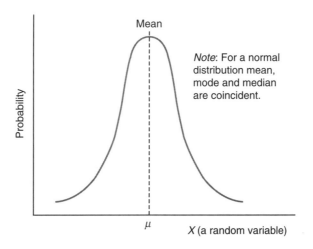

Fig. 5.3   Normal distribution (probability density function).

As, in practice, samples are used, *the standard deviation* of the population must be estimated from that of the sample:

$$\text{Standard deviation of samples} = \sqrt{\frac{\sum(x_i - \bar{x})^2}{n-1}}$$

For samples of $n \geq 32$, it is acceptable to use n rather than $n - 1$.

In many instances, the standard deviation is employed to measure the dispersion as it is on the same scale of measurement as the mean; hence the standard deviation is a more convenient measurement than the variance.

Because the normal distribution is described by a particular formula, from knowledge of the mean and standard deviation, the areas under the curve are known too:

- the range bounded by the curve between $x = \mu \pm \sigma$ contains approximately 68% of the area under the curve,
- the range bounded by the curve between $x = \mu \pm 2\sigma$ contains approximately 95.5% of the area under the curve,
- the range bounded by the curve between $x = \mu \pm 3\sigma$ contains approximately 99.7% of the area under the curve.

Measurements of the spread of a distribution which are simpler than the standard deviation are:

- *Range* – susceptible to distortion by extreme values ('outliers').
- *Interquartile range* – the range between the first and third quartiles (i.e. 50% range of the distribution).
- *Decile range* – the range of the distribution excluding the lowest and highest 10% of data points (i.e. 80% range of the distribution).

The interquartile and decile ranges use the median as the centre of the distribution.

Given a particular value from a sample, the level of confidence concerning its variability can be calculated *provided* it can be established that it is reasonable to say that the normal distribution is applicable and, therefore, parametric tests are appropriate. (Parametric statistics depend upon particular, specified distributions applying to the sample or population for the tests to be valid. Thus, non-parametric statistics provide more powerful tests.)

To calculate the level of confidence, the *standard normal* variable, z is employed:

$$z = \frac{\bar{x} - \mu}{\sigma}$$

The usual levels of confidence are:

- 95% confidence is described as significant (5% significance)
- 99% confidence is described as highly significant (1% significance)
- 99.9% confidence is described as very highly significant (0.1% significance).

Confidence refers to the level of probability that the test results are valid (such as accepting a null hypothesis correctly), that the interval estimate does include the population parameter; while the level of significance (1 – the level of confidence) denotes the probability that the null hypothesis is correct but has been rejected incorrectly (type 1 error).

Null hypotheses

Another form of hypothesis which is encountered in statistical testing is the *null hypothesis*. Commonly, null hypotheses are employed in supplementing the overall hypothesis and any sub-hypotheses and they lend rigour to statistically testing particular, possible relationships between variables. A null hypothesis is tested in comparison with its opposite, the alternative hypothesis. The usual form is that the null hypothesis speculates that there is no difference between, say, the cost in $ per kilometre of travel by coach and by train. The alternative hypothesis speculates that there is a difference. Thus:

$$H_O: \mu_1 = \mu_2$$

$$H_A: \mu_1 \neq \mu_2$$

where: μ_1 = population mean$_1$
μ_2 = population mean$_2$
H_O = null hypothesis
H_A = alternative hypothesis.

Similarly, the approach can be used to examine the value of the mean of a sample and that of the population.

$$H_O: \mu = \$\alpha \text{ per km}$$

$$H_A: \mu = \$\alpha \text{ per km}$$

$$\alpha = \text{constant}$$

To test H_O and H_A, the *t*-distribution is used as, commonly, $n \geq 30$ and σ is unknown. The calculation of t will be compared with the value obtained from the table of the t-distribution.

$$t(\text{calc}) = \frac{\bar{x} - \mu}{s/\sqrt{n}}$$

where: $\bar{x}$ = sample mean
μ = population mean
s = sample standard deviation
n = number in sample.

The level of confidence (or significance) must be decided in order to determine the appropriate value of t from the table of the t-distribution (t(tab)). The degrees of freedom must be calculated to select the correct t(tab). For a t-distribution, there are $(n - 1)$ degrees of freedom to determine t(tab). So, using a two-tailed test, if t(tab) $> t$(calc), the null hypothesis is accepted (i.e. $\mu = \$\alpha$ per km at the selected level of confidence). However, this result does not substantiate that the null hypothesis is true, it shows that there is no statistical evidence to reject it.

To consider H_O: $\mu_1 = \mu_2$ against H_A: $\mu_1 \neq \mu_2$, again the appropriate level of confidence/significance must be decided (commonly 5%). Essentially, the test concerns whether the means of the samples $\bar{x}_1$ and $\bar{x}_2$ are drawn from populations with the same mean. The means of the populations are unknown. The method of testing requires the sampling distribution of the statistic $(\bar{x}_1 - \bar{x}_2)$ to be constructed, this is a normal distribution (Yeomans 1968), the mean of which is zero when $\mu_1 = \mu_2$. The method requires all possible means of each sample distribution to be calculated. All possible pairs of the sample means are considered, one mean being subtracted from the other (i.e. $\bar{x}_1 - \bar{x}_2$). Then, if the mean of the resultant sampling distribution $(\bar{x}_1 - \bar{x}_2)$ is zero, H_O is accepted.

A particular feature of null hypotheses is whether a 'one-tailed' or a 'two-tailed' test is required. If the null hypothesis is of the form H_O: $\mu_1 = \mu_2$, a two-tailed test is necessary where the alternative hypothesis considers $\neq$, in which both greater than or less than values are considered. However, if the alternative hypothesis is of the form H_1: $\mu_1 < \mu_2$, a one-tailed (left-tailed, lower-tailed) test is required (if H_1: $\mu_1 > \mu_2$, a one-tailed, right-tailed, upper-tailed test is required). For a two-tailed test, the region for rejection of the null hypothesis is in both tails of the distribution whilst in the one-tailed test, the region for rejection lies in the appropriate of either the lower or upper tails.

Sometimes testing a hypothesis yields an incorrect result – such error may occur in two ways. A type 1 error occurs when a null hypothesis is correct but the decision taken, based on a test, is to reject H_O, i.e. H_O is rejected incorrectly. A type 2 error is when H_A is correct (H_O is, therefore, incorrect) but the result of testing leads to the decision to reject H_A, and accept H_O, incorrectly. Conventionally, the risk of making a type 1 error is α and the risk of making a type 2 error is β. Ideally $\alpha = \beta = 0$ but, as most data are obtained through sampling (rather than measuring the entire population), the practical objective is to minimise α and β and to keep them small.

The significance level chosen for testing a hypothesis is α, the probability of making a type 1 error. Depending on the consequences of making a wrong decision (from consideration of the null and alternative hypotheses), a decision maker will be able to say which type of error is more onerous – i.e. which type of error is worse; if it is a type 1 error (falsely rejecting a null hypothesis which is correct) a high level of significance (α at 1% or 0.1%) should be chosen. The question which remains is, 'what is the size of β'?

The value of $(1 - \beta)$ measures the probability of rejecting a null hypothesis when it is false; this is the objective of testing null hypotheses. Hence, high values of $(1 - \beta)$ are desirable. $(1 - \beta)$ is called the power of the test so, the smaller the value of β, the greater is $(1 - \beta)$ and, therefore, the power of the test. As changes in the sizes of α and β work inversely (as a trade-off) if α is reduced, β will increase.

Validities

Frequently, research is concerned with investigating a hypothesised causal relationship between an independent variable and a dependent variable; if such a relationship is found, inferences are drawn about the population and, perhaps, a variety of circumstances in which the relationship may apply beyond those of the particular study carried out. Such research involves a set of *validities* (the likely truth of a hypothesis). They are described as construct, internal, statistical inference and external validities. (See also: experimental design – Chapter 4; obtaining data – Chapter 6.)

Construct validity concerns the degree to which the variables, as measured by the research, reflect the hypothesised construct. Poor construct validity occurs if the measurements are caused by other variables' influence or random noise. Bacharach (1989) warns that 'embedded in the construct validity concept is the assumption that the variables were correctly measured.'

Internal validity is high where the observed and measured effect is due to the identified causal relationship; i.e. changes in the independent variable(s) are the sole (main) cause of changes in the dependent variable. To achieve good internal validity, care is needed in the research design such that alternative explanations are examined and appropriate methods selected by which the causality can be investigated.

Errors due to inadequate internal validity generally arise from the presence of one or more common threats:

- *History* – have any relevant and (potentially) significant data been omitted from the study?
- *Instruments* – are measures accurate and reliable, especially if made several times?
- *Maturation* – would the changes measured have occurred in any case/ naturally, rather than being due to the attributed cause?
- *Mortality* – has a particular and significant withdrawal of (a) data source(s) occurred?
- *Regression* – if measures are repeated, low/high scores naturally tend to move towards the mean.
- *Selection* – has there been any bias in the sources of data selected?
- *Testing* – ('Hawthorne effect') were the results caused by the research process itself?

In comparative studies, especially if longitudinal, further threats to internal validity involve *diffusion*, where respondents/participants (notably a control group) become familiar with the research process/questions etc. and this may influence the data obtained. Rivalry may arise or people may become demoralised; this may apply to the researchers as well as to the respondents/participants.

Generally, as internal validity is increased and, hence, confidence in the accuracy of the results etc. rises, the level of generalisability – the scope of external validity – is reduced.

Statistical inference validity, judged by inference statistical measurements, is high where the sample is a good representation of the population. Hence, effects on the population can be inferred with a high level of confidence from the behaviours of the sample because the statistics of the sample are close approximations to the parameters of the population.

External validity concerns the degree to which the findings can be generalised over circumstances which are different from those of the tests carried out; the initial concern of external validity is the applicability of the results, drawn from testing a sample, to the population. It concerns the questions of how restrictive is the study; are the findings applicable to other populations? Judgement of external validity requires careful comparison of the sample and the population from which it was drawn with other populations. This should include a comparison of circumstances for the two populations.

Threats to external validity are:

- Interaction of selection and treatment which prevents valid generalisation of the findings beyond the sample studied.
- Interaction of setting and treatment which prevents generalising beyond the particular setting.
- Interaction of history and treatment which means that the results are time-dependent.

For in-depth discussion of validities, refer to Dooley (1990).

Jung (1991) discusses a possible inverse relationship between the internal and external validities of experiments – the 'experimenter's dilemma' – high internal validity tends to produce low external validity and vice versa. Commonly, external validity is not a crucial consideration, although it may be important (Mook 1983); the results of a particular study often depend too greatly on the context to facilitate generalisation of the findings, but internal validity is essential. In some cases, external validity is important, for instance, surveys which aim to predict from a sample to the population. Frequently, in experimental research, the purpose is to test hypotheses about the effects of certain variables – experiments are designed to determine whether, and to what extent, the hypotheses are supported by the data. If the evidence supports the hypotheses and, hence, the theory, the results may be used to generalise the theory, not the results themselves, in other contexts. Generalisability of a theory is determined through replicating experiments in different contexts, with different subjects and, further, by using modified experimental techniques and procedures.

Cordaro and Ison (1963) consider the possible effects of the experimenter expecting particular results and the resultant consequences. Usually, researchers have an idea or expectation of how subjects will respond to questions, or how materials in laboratory experiments will behave etc. – frequently such expectation, or its opposite, provides the explicit hypothesis for the research. Thus, it is important to be aware of potential bias in that such 'expectations' can distort the results by affecting how the researchers interpret data such as responses etc. To help to counteract any such bias, the researcher might adopt an hypothesis which is the opposite to the researcher's expectations.

Summary

This chapter has considered the definition of an hypothesis and proceeded to examine how appropriate it is to formulate hypotheses for different types of studies. Where appropriate, hypotheses aid both objectivity and delineation of the parameters of the research. The role of sampling was discussed and basic statistical measures for means and dispersion were noted. The use of null hypotheses was examined and how such hypotheses assist determination of the level of confidence in research findings. Common types of error in null hypotheses testing were discussed.

References

Bacharach, S.B. (1989) Organizational theories: some criteria for evaluation, *Academy of Management Review*, **14**(4), 496–515.

Cordaro, L. & Ison, J.R. (1963) Psychology of the scientist: X, observer bias in classical conditioning of the planaria, *Psychological Reports*, **13**, 787–789.

Dooley, D. (1990) *Social Research Methods*, 2nd edn, Prentice-Hall, Englewood Cliffs, NJ.

Jung, J. (1991) *The Experimenter's Dilemma*, Harper & Row, New York.

Mook, D.G. (1983) In defense of external invalidity, *American Psychologist*, **38**, 379–387.

Moser, C.A. & Kalton, G. (1971) *Survey Methods in Social Investigation*, 2nd edn, Dartmouth, Aldershot.

Oxford English Dictionary (2007) http://dictionary.oed.com.eproxy1.lib.hku.hk/cgi/entry/00315594?, accessed 10 July 2007.

Popper, K.R. (1989) *Conjectures and Refutations: The Growth of Scientific Knowledge*, Routledge & Kegan Paul, London.

Schick, T. & Vaughn, L. (2002) *How to Think about Weird Things: Critical Thinking for a New Age*, 3rd edn, McGraw-Hill, Boston, MA.

Whetten, D.A. (1989) What constitutes a theoretical contribution?, *Academy of Management Review*, **14**(4), 490–495.

Yeomans, K.A. (1968) *Statistics for the Social Scientist 2: Applied Statistics*, Penguin, Harmondsworth.

Chapter 6

Data Collection

The objectives of this chapter are to:

- consider how *data requirements* may be determined;
- discuss issues of *collecting data*;
- identify issues of *sampling* and *sample size*;
- discuss *scales of measurement* and scaling techniques;
- examine problems of *obtaining data* from respondents.

Data requirements

At an early stage of a research project, it is a good discipline to give preliminary consideration to data requirements. For any study which extends beyond a review of theory and literature, a major issue is the collection of data. However, just because a researcher wishes to collect certain data does not ensure that those data will be available. Restrictions on collection of data apply for a variety of reasons – confidentiality, ease of collection or provision, cost, time etc.

Despite the potential problems, it is helpful to determine what data are ideally required for the research, and then to modify those requirements, if necessary, to overcome practical difficulties. The objective is to obtain an appropriate set of data which will permit the research to proceed to test any hypothesis and to realise the objectives (to address the research question(s)) as rigorously as possible, given the dynamism of research and the practical considerations, with outputs reasonably close to the original intentions.

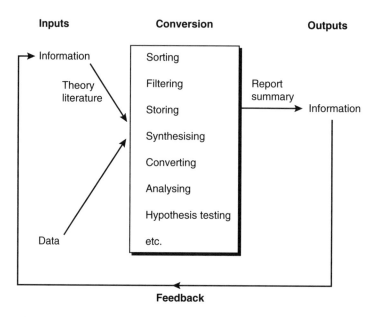

Fig. 6.1   Simple model of an information system.

Essentially, a research project is a form of information system. Fig. 6.1 is a simple model of an information system. For a discussion of information systems in construction, see, for example Newcombe *et al.* (1990, pp. 115–135).

To determine the inputs for an information system, the outputs required must be decided first. After consideration of the conversion process, deciding on the analyses etc. which will be carried out to yield the outputs, the input requirements can be determined. Note that the system is analysed from outputs to inputs. For a research project, start with what outputs are desired (the questions to be answered, objectives to be met, hypotheses to be tested), then consider what analyses can be carried out and what alternatives are appropriate. From this one can decide the data (input) requirements. However, the very nature of research, as a 'voyage of discovery', dictates that the outputs will not be known until the research has been done, and may not be known even then.

The research outputs should be considered in terms of the aim, objectives and hypothesis, where relevant, plus sub-hypotheses. So, whilst the outputs in terms of results, findings and conclusions are unknown, the issues which they concern have been determined to a large extent. Such information is available irrespective of the nature of the research project; whether it be a qualitative study involving fundamental investigation of, say, behaviour in a novel environment, where an aim and some objectives will have been established, or a quantitative study, with strict adherence to scientific method, in which aim, objectives and hypothesis will have been formulated at an early stage.

In systems theory, *bounded* systems tend to be 'closed' and, as such, they are isolated from their environment by an impenetrable boundary. In this sense, quantitative studies are bounded, often rigidly, as the variables have been identified and the primary task is to test the hypothesis which has been formulated at the outset of the study or after the review of theory and literature but which remains a precursor to the empirical work. 'Open' systems are *unbounded*, they have a highly permeable boundary, and so the system acts with, and in response to, changes in the environment. Qualitative studies are relatively unbounded as, whilst pursuing investigation of aim and objectives, they collect all possible data to detect variables and relationships etc. Boundaries are important both in their location and their nature – see Figs. 6.2(a) and (b). A good discussion of the systems theory approach is contained in Cleland and King (1983).

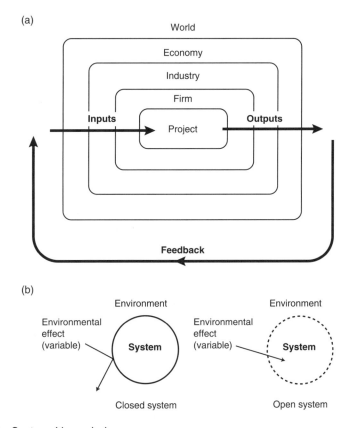

Fig. 6.2   Systems' boundaries.

Collecting data from respondents

Collection of data is a communication process. Not only may it involve transfer of the data from the provider (respondent) to the collector (researcher), it
may also involve the provider in collection, assembly etc. of data. Thus, the
collection may involve a chain of communication – much of which may be
invisible to the researcher. That aspect merits investigation, recording and,
if possible, checking/auditing to ensure the reliability and accuracy of the
data obtained; this can be a critical factor where triangulation of data yields
differences in results in seeking explanation of such differences.

The primary aim in collecting data is to maximise the amount and accuracy of transfer of meaning (convergence) from the provider to the researcher.
Much of the likelihood of success of convergence is determined by the methods selected for data collection and the expertise with which the methods
are employed. In structured methods, the pre-determination of what data are
to be collected is critical. Selection of providers is also likely to be critical,
although the statistical principles of (large) sampling endeavour to overcome
such criticality by ensuring that the sample is a good representation of the
population. For experiments, case studies etc., there is recognition of the indi-
vidualities of the data etc. – analytic generalisations (from theory rather than
statistical inferences) are used.

Methods of collecting data, generally, may be categorised as either one-way
or two-way communications. One-way methods require either acceptance of
the data provided or their rejection, clarification, checking etc. are possible
only rarely. One-way communication methods include postal questionnaires,
completely structured interviews, diaries, scrutiny of archives/documents
and observations by the researchers. Two-way methods permit feedback
and gathering of further data via probing and include semi-structured
interviews and participant observation. Totally unstructured interviews
are virtually one-way communication. Usefully, one-way communication
methods may be regarded as linear data collection methods whilst two-
way communications methods are non-linear. Fig. 6.3 shows the spectrum
of types of interview related to the nature of the questions which may
be used.

Rogers and Kincaid (1981) assert that linear methods focus on transfer
of data/information whilst non-linear methods are more conducive to the
transfer of meaning. This is explained by linear methods failing to provide
interaction in data collection and so, tending to be used for cross-sectional,
one-off approaches. Interaction – notably, feedback and checking – is very
important in researching human behaviour etc. to ensure that the message
(data) obtained from the providers is understood as per the providers' view(s),
rather than from the perspective of the researcher alone – (i.e. avoiding

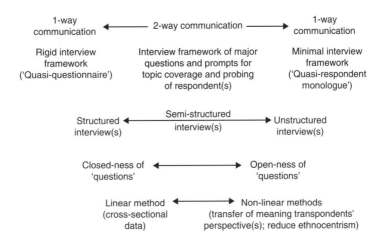

Fig. 6.3   Types of interview.

ethnocentrism). However, in practice, some level of ethnocentrism may be unavoidable – the aim is to render its presence insignificant.

Much research is designed to investigate the cause(s) of events/relationships. The initial step is to determine whether a relationship exists and, if so, whether that relationship is statistically significant and, then, if so, at what level of statistical significance (confidence).

Drenth (1998, p. 23) asserts that three further conditions must be met in order to establish causality:

> 'the supposed cause–effect relationship must be theoretically plausible; the relationship must not disappear when a third variable is introduced into the analysis; [and] the causal variable must precede the effect variable.'

In field studies and field (quasi-) experiments, many unwanted, external effects are likely to impact on the data and it will be impossible to control them or, even if identified, allow for them adequately in the analysis. Even in comparative studies, no two projects, firms, departments etc. are really the same. No 'control group' can really be established and used properly. Hence, there are many and large threats to internal validity.

Drenth *et al.* (1998) discuss the analysis of different designs of quasi-experiments given in Cook and Campbell (1976, 1979) as noted in Table 6.1.

Surveys

The essence of a survey is that it produces information about an entire population. However, because populations tend to be far too large for a full

Table 6.1   Quasi-experiment models.

Design	Process	Notes
(1) One group, post-test only	XO	Unacceptable. No measure of any change possible, so no evidence of an causality
(2) One group, pre-test and post-test	O_1XO_2	Inadequate. No evidence of cause(s) of any change; change could be due to maturation etc.
(3) Untreated control, group with pre-test and post-test	O_1XO_2 / O_1O_2	Care essential to select comparable groups (i.e. to maximize internal validity)
(4) Reversed treatment, non-equivalent control group with pre-test and post-test	$O_1X \uparrow O_2$ / $O_1X \downarrow O_2$	As (3), changes between O_1 and O_2 should be in opposite directions for the two groups
(5) One group, removed treatment with pre-test and post-test	$O_1X + O_2$ / $O_3X - O_4$	The treatment added at X+ is removed at X−; if the change O_1 to O_2 is reversed at O_3 and O_4 then that evidences X to be causal. O_2 and O_3 do not need to be equal, nor do the magnitudes of the changes but they should be similar in size and reversed in direction
(6) One group, repeated treatment with pre-test and post-test	$O_1X + O_2X - O_3X + O_4$	Most interpretable when O_3 and O_4 are different from O_1 and O_2

Key: Experiment X, observation O.

Source: Cook and Campbell 1976, 1979.

population survey to be possible, it is usual for surveys to employ sampling such that the size and structure of the sample is sufficient to yield enough reliable data for inferences to be drawn about the population *at a required and specified level of confidence*. The objective is statistical validity.

Much research in the social sciences and management spheres involves asking and obtaining answers to questions through conducting surveys of people by using questionnaires, interviews and, sometimes, case studies. Often, responses are compared to 'hard data', such as the cost records of a project. However, many 'hard data' are not totally 'objective' or reliable in the sense of showing what they may be believed to show.

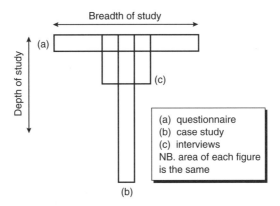

Fig. 6.4   Breadth v. depth in 'question-based' studies.

Example
Commonly, project costs, known as the *out-turn* costs (such as a project final account), are influenced by the negotiating powers and skills of the project participants and so, whilst quantifying cost, a factor of human objectives and skills is incorporated in the final cost figures which result.

Given that there is a finite amount of resources available for carrying out the field work, especially where those resources are very restricted, a choice of research method is necessary. The choice is affected by consideration of the scope and depth required. The choice is between a broad but shallow study at one extreme and a narrow and deep study at the other, or an intermediate position – as shown in Fig. 6.4.

Survey techniques, such as questionnaires, interviews etc., are highly labour intensive on the part of respondents and particularly on the part of the researcher; one consequence is the low response rate, which is common, notably for postal questionnaires which can expect a 25–35% useable response rate. Thus, many surveys do not produce data from which results capable of strong support for tests of hypotheses or conclusions can be drawn. Further, self-completed responses are very prone to bias and distortions – giving answers which respondents believe 'should' be given rather than providing their 'true' answers; giving answers to 'please' the researcher etc. – as well as being self-perceptions by the respondents.

Questionnaires

Questions occur in two primary forms – open or closed. Open questions are designed to enable the respondent to answer in full; to reply in whatever form, with whatever content and to whatever extent the respondent wishes (in interviews, the researcher may probe). Such questions are easy to ask but may be difficult to answer, the answer may never be full/complete and, often, the answers are very difficult to analyse. It is essential that answers to open questions are recorded in detail and in full – in interviews, at least audio recording of the responses is advisable. Closed questions have a set number of responses as determined by the researcher. However, such rigidity of available responses may constrain the responses artificially, hence a response opportunity of 'other, please state' should be provided wherever possible. Care must be taken that responses to open questions are not biased by the response alternatives provided by related and preceding closed questions. Thus, it may be preferable to place open questions before related, closed questions. It is possible to ask more closed than open questions, as responses to closed questions can be given more easily and quickly.

Filter questions may be employed to progress certain respondents past a set of questions which are not relevant to them. Although the technique speeds respondents through the survey and maintains relevance of the questions answered, extensive use of filter questions can be annoying. Funnelling can be used where issues are introduced by a general question and pursued by questions of increasing detail on aspects of the issue being researched. Especially for attitude studies, a form of funnel-sequence of questions is likely to be helpful:

Example
(1) Questions to discover the respondents' awareness/knowledge/thoughts about the issue.
(2) Questions to discover the respondents' general views of the issue (open questions).
(3) Questions concerning specific facets of the issue being researched (closed questions).
(4) Questions to discover reasons for the views held by the respondents.
(5) Questions to discover how strongly the respondents hold their views.

Questionnaires may be administered by post or email/web to respondents, to groups by the researcher or particular individuals, such as to a class of students, by a lecturer, or to individuals by the researcher – perhaps to form the basis of an interview. The questions should be unambiguous and easy for the respondent to answer, they should not require extensive data gathering by the respondent. They should not contain requests for unnecessary data, for instance, they should not request a name when the respondent is known, if the questionnaire was sent to the person by name, especially when anonymity is to be provided or when the identity of the respondent is not needed. Questions should be clear, each should concern one issue only and the request for answers should be given in an 'unthreatening' form appropriate to the research.

Example
Seek a respondent's age by requesting that they tick the appropriate 10-year age band, ensuring the bands do not overlap – e.g. 10–19, 20–29 etc.

Questions may seek opinions on an issue by degree of agreement with a statement, phrased objectively. It may be preferable to use a 4-point rather than a 3- or 5-point scale for the answer (strongly agree, agree, disagree, strongly disagree) to avoid any 'opting out' or indecision by respondents' selecting the middle, non-committing answer. However, odd number response scales do permit genuine 'centrality' to be represented accurately. (There is no 'correct' choice of the number of points to use in a scale – odd or even – both have *pros* and *cons*.)

Questions concerning ordering of hierarchies of issues, derived from literature, should not contain too many items to be ranked; such questions should include the opportunity for respondents to note their own items which do not appear on the list – include an item of 'other, please note'. A useful alternative is to ask respondents to rank their top 5 or 6 items. Rating scales tend to be preferable to ranking scales as they can accommodate more items in the scale, employ appropriate detail of ratings and the results can yield rankings of the items also; however, care must be taken to ensure that the method used to produce the ranking from the rating is appropriate for the nature of the rating scale employed (see Chapters 6 and 7).

If a statement from the respondent in answer to a question is requested, ensure that the statement required can be provided in a few words (the space provided on the form will suggest the size of answer required). Long statements are tedious for respondents and may be difficult to analyse.

Questions concern facts, knowledge and opinion. It is important to appreciate that people's memories are imperfect. Whilst facts may be checked in some cases, and a respondent's knowledge level can be assessed, opinions must be taken at their 'face value' which may be problematic in instances where opinions expressed are inconsistent.

All questionnaires should be piloted initially; completed by a small sample of respondents. The piloting will test whether the questions are intelligible, easy to answer, unambiguous etc., and, through obtaining feedback from these respondents, there will be an opportunity for improving the questionnaire, filling in gaps and determining the time required for, and ease of, completing the exercise. Discussion of the questionnaire with the supervisor and other researchers is a useful supplement to the piloting, as it provides a research-oriented view of the questions, the components and assembly of the questionnaire and probable approaches to the analysis of responses. An important aspect of piloting, which is overlooked all too often, is whether the data yielded by the questionnaire (or other data collection instrument) is suitable for analysis (as intended) and, via the analysis, is adequate to give results which facilitate valid testing of any hypothesis and realising the objectives – i.e. can the research question(s) be answered?

Interviews

In seeking to research other persons' 'worlds' (their views, behaviour, etc.), 'we start with the experiencing person and try to share his or her subjective view. Our task is objective in the sense that we try to describe it with depth and detail. In doing so we try to represent the person's view fairly and to portray it as consistent with his or her meanings' (Charmaz 1995, p. 54).

The statement emphasises the desire to be objective, but, importantly, acknowledges that the objectivity is likely to be limited as the others' worlds are subject to perceptions, interpretations, etc. by the researcher in producing the in-depth and detailed description. 'Research cannot provide the mirror reflection of the social world that positivists strive for, but it may provide access to the meanings people attribute to their experiences and social world' (Miller & Glassner 1997, p. 100).

Subjective interpretation occurs throughout the total research process, including the final step of reading the research report.

A major consequence of acknowledging the subjective components of research is the need to make the subjectivity visible and, hence, subject to scrutiny and analysis. This is achieved by scrupulous recording of all aspects at each stage of the research – the 'what', 'why', 'how', 'when' (etc.) aspects of the entire work. That perspective may be manifested not only in the content of the presentation of the research but also in the language and style of presentation used. It is helpful if the researcher is portrayed as a real and, therefore, fallible person with desires, preferences and prejudices (as we all have), rather than as an anonymous, invisible authority – the 'unquestionable expert'.

Holstein and Gubrium (1997, pp. 113–114) echo Garfinkel (1967) by asserting that 'all knowledge is created from the actions undertaken to obtain it'. Thus, in respect of interviewing, Baker (1997, p. 131) notes that '... interviewing is ... an interactional event ... questions are a central part of the data and cannot be viewed as neutral invitations to speak ... they shape how and as a member of which categories the respondents should speak, ... interview responses are treated more as accounts rather than reports ...'. In order to clarify and to secure categorisation of respondents from their perspectives, rather than the perspectives of the researchers, Baker (1997, p. 136) advises that the interviewer '... asks respondents to reveal, describe, report on their interiors, or on their external world as they know it'.

Interviews vary in their nature, they can be:

- structured,
- semi-structured and
- unstructured.

The major differences lie in the constraints placed on the respondent and the interviewer. In a structured interview, the interviewer administers a questionnaire, perhaps by asking the questions and recording the responses, with little scope for probing those responses by asking supplementary questions to obtain more details and to pursue new and interesting aspects. In unstructured interviews, at the extreme, the interviewer introduces the topic briefly and then records the replies of the respondent. This may be almost a monologue with some prompts to ensure completion of the statements; clearly the respondent can say what and as much as she/he desires. Semi-structured interviews fill the spectrum between the two extremes. They vary in form quite widely, from a questionnaire-type with some probing, to a list of topic areas on which the respondent's views are recorded.

Whichever type of interview is adopted, Pettigrew (1997) emphasises the essential of a good theoretical base to ensure appropriate data are collected, 'A theoretically informed interview pro-forma is an important mechanism to build structure into the data collection process.'

The inputs of the interviewer are critical – especially probings – as the questions asked, the *probes*, will influence the responses obtained. The non-verbal communications or 'body language' of the participants will have an impact on the responses and recordings. Often, with permission of the respondents, tape recording the interviews can be very helpful at the later stages of analysis and, through subsequent scrutiny, help to ensure accuracy and objectivity in recording responses. Transcribing is lengthy, tedious and expensive so the tape recordings may be employed to supplement the interviewer's notes.

Usually, interviews are carried out with individual respondents one-at-a-time. That approach is quite labour intensive for the researcher and, almost always, excludes (by design) interaction between respondents – which may be required to prevent cross-respondent influences and regression to the mean in responses. An alternative design may seek to elicit general consensus views of, usually expert, respondents. That suggests a group interview research design which allows, or encourages, interaction between respondents so that they may amend their initial responses, in the light of responses of others, and thereby, establish a cohesive consensus view (a desired but not a necessary outcome – individuals may retain diverse views, despite knowledge of the views of peers).

The two best-known group interview methods are the Delphi method and the focus group method. In the Delphi method (technique) an array of experts are interviewed or requested to complete a questionnaire individually, the results are then circulated to all respondents (often as descriptive statistics of aggregate responses to each question – means etc. – for quantified answers) and a second round is then conducted (same questions to the same respondents) to elicit their responses, given the responses of the other experts. Further rounds may follow similarly until respondents' responses are, approximately, constant – indicating expert, consensus view(s) on each question.

'The focus group method is an established rigorous technique for collective interviews aimed at eliciting and exploring in-depth opinions, judgements and evaluations expressed by professionals, experts or users/clients about specific topics.' (Morgan 1997).

Case studies

Often, case studies employ a variety of data collection techniques – archival data, interviews, questionnaires, observations, etc. Unlike questionnaires and interviews when the case researched is the respondent and so a possibly large number of cases are researched for statistical significance, in a case study the case is the particular occurrence of the topic of research. It may be, for instance, a legal case hearing, a building in use over a time, or the procurement of a construction project. Interviews may be used accompanied by collection of 'hard' documentary data. Questionnaires are less usual although they may be employed to gain an understanding of the general situation of which the case being studied is a particular instance. A case study yields deep but narrow results. Commonly, it will employ triangulation both in the case study itself and to facilitate generalisation of findings. However, it is essential to be aware of the validity of generalising the findings of a case study research project.

Triangulation

Triangulation is the use of two or more research methods to investigate the same thing, such as experiment and interviews in a case study project. A postal or other questionnaire to a generalised, representative sample of respondents would assist the researchers to appreciate the general validity of the findings from the particular case study and would serve to aid understanding of its unique and generally applicable features.

Further techniques of data collection involve asking respondents to keep diaries over particular periods and/or for researchers, possibly aided by cameras etc., to make observations, akin to a laboratory notebook in natural science experiments. Whatever methods are adopted, it is important that they are implemented as rigorously as possible to try to avoid bias and to obtain appropriate amounts of accurate data.

It is important to be aware of methodological considerations, the advantages and disadvantages of particular methods, error sources, possible bias, strengths of triangulation etc. in order that the validity of the study and, in particular, its results and conclusions can be appreciated and, perhaps, quantified. Jick (1979) notes that 'The effectiveness of triangulation rests on the premise that the weaknesses of each single method will be compensated by the counterbalancing strengths of another. That is, it is assumed that the multiple and independent measures do not share the same weaknesses or potential bias . . .. Although it has always been observed that each method has assets and liabilities, triangulation purports to exploit the assets and neutralise, rather than compound, the liabilities.'

Sampling

The objective of sampling is to provide a practical means of enabling the data collection and processing components of research to be carried out whilst ensuring that the sample provides a good representation of the population; i.e. the sample is representative. Unfortunately, without a survey of the population, the representativeness of any sample is uncertain, but statistical theory can be used to indicate the representativeness. Measurements of characteristics, such as the mean, of a sample are called statistics whilst those of a population are called parameters. How to obtain representativeness begins with consideration of the population. Almost invariably, it is necessary to obtain data from only a part of the total population with which the research project is concerned; that part of the population is the *sample*. All buildings on Hong Kong island or all buildings in Greater London can be viewed as populations, whilst both of these are also samples of all buildings in the world. So, the context of the research indicates the population of concern.

There are two main problems which seem to be overlooked by many researchers; they are ignoring considerations of sampling error when determining the size of the sample and failing to address possible biases in groups of respondents and non-respondents. That second problem is particularly important for low response survey methods – notably, postal questionnaires – because, despite low response rates (less than 20% is not unusual), inferences are made about the entire population from analyses of the responses received only, sometimes without even acknowledgement of the potential for bias in the results and conclusions.

The first task is to define the population. Eisenhardt (1989) stresses that '... the concept of the population is critical, because the population defines the set of entities from which the research sample is drawn. Selection of an appropriate population controls extraneous variation and helps to define the limits for generalizing the findings.'

If the population is sufficiently small, a full population 'sample' may be researched, but in the vast majority of cases a sample must be taken; the question is how? At this point, it is helpful to recall some standard symbols (or notation):

Population size	$\rightarrow N$;	Sample size	$\rightarrow n$
Population mean	$\rightarrow \mu$;	Sample mean	$\rightarrow \bar{x}$
Population standard deviation	$\rightarrow \sigma$;	Sample standard deviation	$\rightarrow s$

Given an extremely large population (e.g. the population of China), whilst the population is actually finite, for practical considerations of time, cost etc., the population is approaching infinite. If there is no evidence of variations in the population's structure, if there is reason to ignore the structure, or if the sample is sufficiently large *random sampling* will be appropriate. Alternatively, it may be necessary to constrain the sampling – perhaps by considering only one section of the population or/and by reflecting the structure, or a particular structure of that constrained, sampled population. Such 'structured' sampling requires a 'sampling frame' to be established *explicitly*. Within the sampling frame, random sampling, judgemental sampling or *non-random sampling* may be used.

In random sampling, each member of the population has an equal chance of being selected. Usually, such sampling occurs without replacement of the selected members of the population; the members of the population selected are excluded from further re-selection so that each member of the population can be selected only once. The selection of members to be sampled is carried out using random numbers, either from tables or from computer programs; an appropriate way of allocating a unique number to each member of the population must be devised for this sampling technique to be used.

In some instances, judgemental sampling may be used; the judgement, which one hopes will be well informed, of a person is used to determine which items of the population should form the sample. Such an approach may be very useful in pilot surveys. Alternatively, it is common for judgement to be used to determine which sections, strata or clusters of the population will be sampled; clearly, such sampling may introduce bias, which must be recognised and noted clearly – e.g. investigating behaviours of large companies.

Non-random samples are obtained by:

- systematic sampling
- stratified sampling
- cluster sampling.

Systematic sampling does have an element of randomness. Having determined the sample size, every xth member of the population is sampled, where x is the interval between them and is kept constant. Beginning with the yth member of the population, y is selected at random and counted from the first member of the population or some other, appropriate starting point – e.g. people walking past a certain location, beginning at a particular time of day. Given a population of known size, the sampling fraction (n/N) can be used to determine the members of the population to be sampled systematically, where N is the total population and n is the number in the sample. The sampling fraction indicates the members of 'groups' in the population (this will be x). Random numbers provide the starting member of the first group (y), so the members of the samples are $y, y + x, y + 2x, \ldots$. The sampling is in the form of an arithmetic progression (or arithmetic series) which sums to $y + (x^2 - 1)$.

Stratified sampling is appropriate where the population occurs in 'distinct', groups or strata. Strata may be selected by others for their own purposes – such as national, published statistics, e.g. the UK statistics of sizes of firms – measured by strata, numbers of employees etc. Strata may be selected for the purposes of the research by, e.g. type of firm – industrial, commercial, etc. Ideally, the differences within each group or stratum should be small relative to differences between strata from the perspective of the research project. Having determined the strata, sampling occurs, most commonly by considering the relative importance of each stratum in the population and using such weighting to divide n, the sample size, between the strata; the members to be sampled are selected randomly from each stratum.

Cluster sampling, e.g. testing medicine over a wide range of people, is appropriate where a population is divided into groups such that it is likely that inter-group differences are believed to be small, whilst intra-group differences are believed to be large. The population is divided into the groups, called *clusters*, the clusters are selected randomly and the total members of the clusters provide the total sample.

> *Example*
> A supplier of steel reinforcement bars is concerned about the accuracy of the cutting to length of straight bars leaving the factory; the concern relates to all diameters of bars supplied; 6 mm, 8 mm, 10 mm, 12 mm, 16 mm, 20 mm, 25 mm, 32 mm, 40 mm.
>   Depending on the practicalities of the situation, stratified sampling or cluster sampling could be used to examine the accuracy of the lengths of bars.
>
> A stratified sample would examine each stratum of bar by diameter, perhaps in proportion to daily factory throughput, by weight, length or cost. A random sample of bars in each stratum would be measured for accuracy of length. A cluster sample would examine, say, a predetermined number of loads of reinforcement bars, as batched for delivery. Each load would be a cluster, the loads would be selected randomly and would probably each contain bars of a variety of diameters; each load is likely to be somewhat different in the mix of bars it contains. In measuring the lengths of bars in each sampled load, it would be appropriate to note the diameters of the bars measured also.

Two further, non-random means of obtaining samples are *convenience sampling* and *'snowball' sampling*. Convenience sampling may be used where the nature of the research question(s) and the population do not indicate any particular form of sample and so, the researcher collects data from a sample which can be accessed readily (it is convenient). Snowball sampling involves data which are difficult to access, perhaps because the individual sources of data cannot be identified readily. In such situations, the researcher may identify a (very) small number of sources (respondents) and, after collecting data from each one, requests that source to identify further sources thereby progressively building a sufficient sample. Given the way in which the sample is selected, both convenience and snowball sampling may involve significant biases hence, the samples should be scrutinised and, if possible, the data tested for bias.

Although some rules of thumb do exist ('large number' statistics require $n \geq 32$; a usable data set of at least 100 responses is needed for factor analysis), there is often much debate over the appropriate minimum size of the sample data set. A rough guide is to get as many data sets as is practical, bearing in mind the analyses to be done – hence the level of confidence in the results.

Generally, designs of surveys endeavour to minimise the possibility of both a type 1 (α) error (H_O is rejected incorrectly) and of a type 2 (β) error (reject H_A, and accept H_O, incorrectly).

Sample size

A particular issue in sampling is determination of the size of the sample. By sampling, a statistic called an estimator is obtained. Estimators should predict the behaviour of the population as accurately as possible, that is the basis of making statistical inferences about the population from the results obtained from analysing the sample. That is achieved by requiring estimators to have four main properties; they should be:

- Consistent
- Unbiased
- Efficient
- Sufficient.

The variance of a consistent estimator decreases as the sample size increases. The mean of an unbiased estimator approximates to the mean of the population; there is an equal chance of the mean of the estimator being more than or less than the mean of the population. This is described by saying that there is no *systematic error*. Systematic error (often called 'bias') is when the errors assume a regular pattern of under- or over-measurement, by a proportion or an amount; such error should be revealed by checking and can be compensated by an adjustment. Further, analysis of the errors may reveal their sources as well as size and that it may be possible to reduce/eliminate the error. Systematic error should be avoidable. Unsystematic error, or *random error*, is almost inevitable, but its size should be kept to a practical minimum by research design and rigour of execution and checking. An efficient estimator has the minimum variance of all possible estimators drawn from the sample. A sufficient estimator is the one which makes most use of the sample data to estimate the population parameter; in particular, the *mean square error* is minimised. Mean square error is an error measure which is used widely; it is the arithmetic average (mean) of the errors squared. Each error is multiplied by itself, which eliminates the sign of the error terms since the square of a negative number is positive. The result is a 'magnified' measure of the average error of the measurements made. The value of the mean square error of the sample is that it provides an unbiased estimate of the variance of the population (σ^2).

Sample sizes are determined by the confidence level required of the estimator. The unknown mean of a population can be estimated with a predetermined level of confidence as shown below:

- in practical sampling, the sampling distribution of means is a normal distribution; for a large sample $\bar{x} = \mu$, and

- the size of the sample and its standard deviation, s, can be used to estimate the standard error of the distribution, $\sigma_{\bar{x}}$ (Yeomans 1968).

As s is the best estimate of σ:

$$\sigma_{\bar{x}} = \frac{s}{\sqrt{n}} \left[\text{approximates to} \frac{\sigma}{\sqrt{n}} \right]$$

For a normal distribution, the 95% confidence intervals are at $\mu - 1.96\sigma$ and $\mu + 1.96\sigma$. So, if z denotes the confidence level required, the confidence interval is:

$$\underline{\mu} = \bar{x} \pm z \frac{s}{\sqrt{n}}$$

where $\underline{\mu}$ is the upper and lower confidence limits of the estimate of μ. Normally, 95 or 99% confidence levels are used, i.e. $z = 1.96$ or 2.58. So, manipulation of the formula for μ yields the sample size required:

$$n = \frac{z^2 \times s^2}{(\mu - \bar{x})^2}$$

As $\bar{x} - \mu$ is half of the width of the confidence interval required, neither of their individual values are required. However, the degree of precision of the estimate which is acceptable must be decided.

Example

Consider the prices in dollars of a particular type of car. 'Experience' may indicate that a precision of $\pm 5\%$ is appropriate, which for a car costing about $5000 translates into about $\pm\$250$ (i.e. $\mu - \bar{x} = \$250$).

If a sample of cars in the UK indicates that $\bar{x}_{UK} = \$5000$ and $s_{UK} = \$700$, what sample size of similar cars in Australia should be employed if 95% confidence is required and an estimate provision of $\pm\$250$ is acceptable?

$$n = \frac{1.96^2 \times 700^2}{250^2}$$

$$= \frac{2.8416 \times 490\,000}{57\,500} = 24.2$$

So, a sample of 25 cars in Australia would yield an adequate solution. Note the assumption that the standard deviation of the sample is the same in each country (i.e. $S_{UK} = S_A$).

> *Example   Continued*
> An alternative approach to the sampling problem is to determine the probability that the mean of a sample is within a prescribed range of the known mean of the population. So, if the mean price of cars in UK is £9000 with $\sigma = $ £1200, what is the probability that the mean price of a sample of 50 cars will lie between £8700 and £9050?
>
> $$\sigma_{\bar{x}} = \frac{\sigma}{\sqrt{n}} = \frac{1200}{\sqrt{50}}$$
>
> $$= £169.71$$
>
> Using *z*-values to calculate the probability, and given that:
>
> $$z = \frac{\bar{x} - \mu}{\sigma_{\bar{x}}}$$
>
> For the extreme values of $\bar{x}$:
>
> for $\bar{x} = $ £8700.00   for $\bar{x} = $ £9050.00
>
> $$z = \frac{8700 - 9000}{169.71}$$ $$z = \frac{9050 - 9000}{169.71}$$
>
> $$= -1.79$$ $$= 0.30$$
>
> Areas from *z* table are 0.4633 and 0.1179. By adding the areas, the probability that the mean of the sample of 50 cars lies between £8700.00 and £9050.00 is 0.5812.

The *central limit theorem* considers the relationship between the shape of the population distribution and the shape of the sampling distribution of the mean. For non-normal distributions of a population, the sampling distribution of the mean approaches normality rapidly as the size of the sample increases, so the mean of the sampling distribution $\mu_{\bar{x}}$ equals the mean of the population (μ). The central limit theorem, therefore, allows inferences about population parameters to be made from sample statistics validly, without knowledge of the nature of the frequency distribution of the population. Although there are instances where the central limit theorem does not apply, these instances are quite uncommon.

> 'One method of determining sample size is to specify margins of error for the items that are regarded as most vital to the survey. An estimation of the sample size needed is first made separately for each of these important items.'   (Cochran 1977, p. 81)

Hence, an important consideration is which variables are to be included. Once those major variables have been identified, the sample size based on each variable individually can be calculated. Provided the calculations indicate that each sample size is approximately equal, the largest sample should be adopted. However, the results may be quite different from each other or the sample sizes may be impractically large; in the former instance, some variables may have to be eliminated whilst in the latter instance, a reduced level of precision may have to be accepted to allow a smaller sample size to be used.

It is important to remember that the sample size consideration relates to usable data sets and not the size of the sample approached to elicit data (responses). Hence, the results of sample size calculations must be modified to accommodate response rates in determining the size of sample to be approached for data (as indicated below).

Cochran (1977) calculates the sample size for continuous data (e.g. success):

$$n_s = \frac{(t)^2 x(s)^2}{(e)^2}$$

where: n_s is the sample size
          t is the value for alpha (usually 0.05, which yields a t-value of 1.96 for sample size of 120 or more; refer to statistical tables for other values of alpha and for small sample sizes).
          s is the estimated standard deviation of the population;
          e is the acceptable level of error for the mean being estimated (commonly 3% for continuous data and 5% for categorical data).

It is common to measure continuous data on a cardinal, or interval, scale (see below). If a 5-point scale were adopted, then the value of e would be $5 \times 0.03 = 0.15$. Further, given that, for a normal distribution, the mean ±3 standard deviations encapsulates approximately 99.8% of the area under the curve, $s = 5/6$.

If Cochran's initial formula results in a sample size which is more than 5% of the population, then a correction should be applied to arrive at the final sample size:

$$n_f = \frac{n_s}{(1 + n_s/P)}$$

where: n_f is the final sample size
          P is the size of the population.
          For categorical data, the sample size is:

$$n_s = \frac{(t)^2 x(p)(q)}{(e)^2}$$

where: (p) (q) is the estimate of the variance (Krejcie and Morgan (1970) suggest $p = q = 0.5$ for dichotomous variables where the population, reasonably, is expected to be divided equally).

Any adjustment to produce n_f is calculated from the formula, above.

For studies in which regression analysis is likely to be employed, the consensus seems to be that the ratio of observations to each independent variable should be at least five, with the more common ratio advocated being ten. The relationship is required to try to avoid the problem of 'overfitting' which makes the results specific to the sample data and so, they lack generalisability (external validity). It must be borne in mind that each dummy variable must be included in the calculations. (A dummy variable is any variable which must be allocated a value of 1 or 0 to indicate whether it is present in the data – yes/no: e.g. gender, 1 = female, 0 = male is a single dummy variable; nationality, if 5 nationalities were being examined, each would be a dummy variable such that 1 = yes, 0 = no for each nationality.)

For studies in which factor analysis is likely to be used, a minimum of 100 observations are required. Increasing the number of observations (sample size) reduces the level at which an item loading on a factor is significant.

Scales of measurement

Often, types of data are identified in terms of the nature of the scales of measurement used. The four primary scales of measurement are *nominal, ordinal, interval* and *ratio*. Measurement simply means assigning numbers or other symbols to characteristics of objects according to set rules; scaling involves the generation of a continuum upon which measure objects are located. Malhotra and Birks (1999, p. 293) provide an example of the four measurement scales as follows.

Nominal scale   This is the most basic scale as the numbers are used only for identifying or classifying objects, e.g. numbers assigned to the marathon runners.

Ordinal scale   The numbers indicate the relative position of the objects but not the magnitude of difference between them, e.g. rank order of the winners in the marathon – 1st, 2nd, 3rd and so on.

Interval scale   Permits a comparison of the differences between the objects, e.g. performance rating (time intervals between the marathon runners) on a 0–10 scale.

Table 6.2   Scales of measurement and data analyses.

Scale	Basic Characteristics	Example	Common Descriptive Statistics Used	Common Inferential Statistics Used
Nominal	Numbers identify and classify objects	Gender classification	Percentages, mode	Chi-square, binomial test
Ordinal	Numbers include the relative positions of the objects but not the magnitude of differences between them	Ranking of quality delivered by a number of construction projects	Percentile, median	Rank-order correlation
Interval	Differences between objects can be compared; zero point is arbitrary	Attitudes, opinions, index numbers	Range, mean, standard deviation	Product–moment correlations, t-test, ANOVA, regression, factor analysis
Ratio	Zero point is fixed; ratios of scale values can be computed	Construction costs, project completion time, market shares	Geometric mean	Coefficient of variation

Ratio scale   The ratio scale possesses all the properties of the nominal, ordinal and interval scales, and, in addition, an absolute zero point. Therefore, one can classify and rank the objects (as in height, weight, age and money), and compare intervals or differences, e.g. time to finish the marathon.

Based on the four primary scales of measurement, suggested statistical analyses are given in Table 6.2.

The essential issue concerning scales is that of:

- uniformity of measurement – of distances between measures, grades or graduations, of the commonality of understanding of the measures
- consistency of measurements.

Nominal or *categorical* scales classify members of the sample, the *responses*, into two or more groups without any implication of distance between the groups. Hence, nominal scales provide a mechanism of classification for responses and respondents (e.g. gender, nationality, ethnicity, size groupings).

Ordinal scales are used to rank responses. Again, there is no indication of distance between scaled points or commonality of scale perceptions by respondents. In essence, it provides a hierarchical ordering (e.g. arranging a group of people into a sequence according to their weight; arranging a sample of buildings into a hierarchy according to aesthetic preferences expressed by a sample of members of the public – each respondent may be requested to rate each building on a 5-point, Likert, scale: 5 = like very much, 4 = like, 3 = indifferent, 2 = dislike, 1 = dislike very much).

Interval or *cardinal* scales employ equal units of measurement for the scale. Interval scales indicate the order of responses and distances between them. Use of an interval scale permits statements about distances between responses to be made, but not about relationships in ratio terms between scores; this occurs as the zero point for the scale is selected as a matter of convenience rather than having some basic, absolute fixity of reference (e.g. scales in common use for measuring temperatures – Fahrenheit; Centigrade).

Ratio scales, however, do have a fixed zero reference point, plus the properties of an interval scale. Hence, conclusions can be drawn about both differences in scores (because intervals on the scale are equal – e.g. 1–3, 3–5, 5–7, 7–9), and the relationship between the score – e.g. 8 = twice 4. For a discussion of scales of measurement for variables, see, e.g. Bryman and Cramer (1994) and, in respect of the (numerical) analyses and validity, see MacRae (1988).

One important observation is that although a variable may be continuous (e.g. time), almost always, it will be measured on a non-continuous scale.

Scaling techniques (non-metric and metric)

Scaling techniques are employed in collecting data from the respondents. The scaling techniques commonly employed in construction research can be classified into non-metric (comparative) and metric (non-comparative) scales and the examples are given below:

Non-comparative (metric) scales:
 Itemised rating scales (e.g. Likert, semantic differential, and Stapel scales)
 Continuous rating scales

Comparative (non-metric) scales:
 Paired comparison
 Rank order

Non-comparative (metric) scales

In non-comparative (metric) scales, each object is scaled independently of the others in the stimulus set. The resulting data are generally assumed to be interval or ratio scaled, e.g. respondents may be asked to evaluate objects on a 1- to 5-point preference scale (1 = not at all preferred, 5 = greatly preferred). Non-comparative (metric) scales can be itemised rating scales or continuous rating scale.

Itemised rating is more widely used in construction management research – its itemised rating scale can be further classified as Likert 5-point scale, semantic differential or Stapel scales. The respondents are provided with a scale that has a brief description (and, usually, a number – for the extremes, at least) associated with each category. The categories are ordered in terms of scale position and the respondents are required to select the specified category that best describes the object being rated. For instance, a Likert scale usually is applied where each statement is assigned a numerical score from 1 to 5; a semantic differential scale is similar to Likert but goes up to 7 points of rating with each end point associated with bipolar labels that have opposing semantic meanings; a Stapel scale is a unipolar rating scale, usually presented vertically, with 10 categories numbered from -5 to $+5$ without a neutral point (zero).

A continuous rating scale is also referred to as a graphic rating scale, where respondents rate the objects by placing a mark (such as putting a 'tick') at the appropriate position on a line that runs from one extreme of the criterion variable to the other (such as from 0 to 100 on a horizontal line).

Comparative scales (non-metric) scales

Comparative scales (non-metric) involve the direct comparison of stimulus objects, e.g. respondents may be asked if they prefer one object over another. Comparative (non-metric) scales include paired comparisons, rank order, constant sum scales etc. In *paired comparison* scaling, the respondent is presented with two objects and asked to select one according to some criterion – the data obtained are ordinal in nature. It is possible to convert paired comparison data to a rank order. For instance, the researcher can calculate the percentage of respondents who prefer one stimulus over another. In *rank-order* scaling, the respondents are presented with several objects simultaneously and asked to order or rank them according to some criterion. Typically, these data are obtained by asking the respondents to assign a rank of 1 to the most preferred object, 2 to the second most preferred and so on. Rank-order scaling results in ordinal data. An example of rank-order scaling is Conjoint Analysis (see Malhotra & Birks 1999).

Common scaling methods

Perhaps the most common scale for obtaining respondents' opinions is the Likert scale. As noted by Bell (1993), such scales are concerned with determining respondents' degrees of agreement or disagreement with a statement on, usually, a 5- or 7-point scale. By using an odd number of response points, respondents may be tempted to 'opt out' of answering by selecting the midpoint. Hence, it may be helpful not only to keep the number of response points small but also to use an even number of response points, thereby having no central point. So, a 4- or 6-point scale of responses may be preferable to the more usual 5 or 7 points.

The next issue concerns what to do with such a scale of responses; the appreciation of the nature of the scale and hence, what can be determined. As a Likert scale is an ordinal scale (it is a rating scale), it can be used to produce hierarchies of preferences (ranking scales) which then can be compared across groups of respondents as per the sampling frame. Using such an approach, it is possible to determine various groups of respondents' views of an issue by asking respondents from each group to respond to a common set of statements against the Likert scale.

Example
Typical set of statements from which respondents may be asked to choose.

Either: A	(1) Essential	(2) Useful requirement	(3) Helpful addition	(4) Irrelevant

or

B	Strongly agree	Agree	Disagree	Strongly disagree

For **B**, a 5-point scale would employ:

(1) Strongly agree	(2) Agree	(3) Unsure/Uncertain	(4) Disagree	(5) Strongly disagree

Analysis of data from such scales is considered in Chapter 7; in particular, note the example which uses Spearman's coefficient of rank correlation.

Osgood *et al.* (1957) developed a technique using semantic differentials. The technique is suitable for examining the meaning of concepts. Normally, a 7-point scale is used, with only the extremes' being described by adjectives; the adjectives are polar opposites of each other and different adjectives are used for each dimension to be measured. Positions on the dimensions marked by respondents are scored 1–7. To avoid response set bias, it is advisable to vary the positions of the adjectives at the poles (e.g. not all 'good' on the right). The approach requires the researcher to determine the issues to be investigated, the attributes or concepts which are important and suitable adjectives to describe the extremes of each concept's dimension.

The analytic technique for the scores (Moser & Kalton 1971) is to compare means of the respondents' scores for each subject researched, such that respondents' profiles of each subject with respect to the attributes judged to be important are obtained. Factor or cluster analysis would yield more detail. As respondents are presented with scales, the poles of which are identified and described, they note their response on each attribute dimension relative to the poles. Hence, provided it is safe to assume that the understanding of the poles is common amongst the respondents, and that they understand that their position on a continuum between those poles is required, it is appropriate to employ mean scores and other descriptive statistics in preliminary analysis.

> *Example*
> Fiedler (1967) used semantic differentials in his research into leadership – he employed a questionnaire of semantic differentials to produce a Least Preferred Co-worker (LPC) scale. Scoring, using demarcation levels and means, was used to assist examination of managerial styles. (See, e.g. Rowlinson *et al.* 1993.) This method has been replicated, with criticisms.

For measurements such as those described in the above example, the use of means exclusively, with no measures of variability such as variance or standard deviation, may be restrictive to the extent of unreasonableness, thereby rendering the results of dubious use and validity. Median, modal or an index measurements might be more appropriate – see, e.g. Ahmad and Minkarah (1988), Shash (1993) for indexes and scales with Likert scale data. However, Likert scales are ordinal scales and so, although used rather frequently, care must be exercised in using means etc. which are measurements valid if a ratio, or interval, scale is used. For Likert scales to be used to produce means etc. for analyses and interpretation, both the descriptors and their numerical coding counterparts should be noted on the scales, respondents should be told of the assumed equivalence of the descriptors and numerical scores and that the intervals between points on the scale are of (approximately) equal distance, finally, a large number of responses should be obtained (sampling yields more than 32 usable responses).

Development of multi-item scales

The development of multi-item rating scales requires considerable technical expertise and robust methodology in determining the sequence of operations needed to construct the multi-item scales. It must be emphasised that this is not about the 'technical know-how' in formulating the data collection and data analysis to produce a scale (as in BENCHMARKING); rather, the methodology and philosophy of the research approach must be addressed. The theoretical and operational constructs of all the variables, that constitute the theoretical model whereby the scales of measurements of the variables are derived, must be thoroughly examined and scrutinised.

The theoretical and operational constructs dictate the characteristics of the variables to be measured. Scale development begins with an underlying theory of the constructs being measured. Theory is necessary not only for constructing the scale but also for interpreting the resulting scores.

> *Example*
> A multi-item scale (such as KPI – key performance index) should be evaluated for accuracy and applicability which involves an assessment of reliability, validity and generalisability of the scale. Approaches to assessing reliability include, at least, test–re-rest reliability, alternative-form reliability and internal consistency reliability. Validity can be assessed by examining content validity, criterion validity and construct validity.

Reliability refers to the extent to which a scale produces consistent results if repeated measurements are made and is assessed by determining the proposition of systematic variation in a scale. Approaches for assessing reliability include the test–re-test, alternative forms and internal consistency methods.

In *test–re-test reliability*, respondents are administered identical sets of scale items at two different times, under as nearly equivalent conditions as possible. The degree of similarity between the two measurements is determined by computing a correlation coefficient; the higher the coefficient, the greater the reliability. The test–re-test approach is usually applied in conjunction with other approaches such as the alternative-forms reliability approach. In *alternative-forms reliability*, two equivalent forms of the scale are constructed. The same respondents are measured at two different times (e.g. by initially using Likert scaled items and then using Stapel scaled items) and the scores from the administration of the alternative scale forms are correlated to assess reliability. The two forms should be equivalent with respect to content, i.e., each scale items should attempt to measure the same items. A low

Fig. 6.5   Development of a multi-item scale.

correlation may reflect either an unreliable scale or non-equivalent forms. *Internal consistency reliability* is used to assess the reliability of a summated scale where several items are summed to form a total score. Each item measures some aspect of the construct measured by the entire scale and the items should be consistent in what they indicate about the construct. This measure of reliability focuses on the internal consistency of the set of items forming the scale (see Fig. 6.5)

In contrast to the scaling techniques discussed above (comparative, non-comparative, multi-scale) which require the respondents to directly evaluate the construct that the researcher believes to comprise the object of study (e.g. the cognitive state of the project respondent's satisfaction), mathematical scaling techniques allow the researcher to infer respondents' evolutions of the constructs of the object of study. These evaluations are inferred from the respondents' overall judgements and two popular mathematically derived scaling techniques are multidimensional scaling and conjoint analysis, which find most application in analysing data relating to respondents' perceptions and preferences.

Example
Conjoint analysis is used to infer mathematical scaling from the participants' overall judgements in project selection decisions (Liu *et al.* 1994) and value engineering decisions (Liu 1997).

Obtaining data

Given the increasing number of research projects, collecting data is becoming progressively more difficult. The people who are targeted as respondents often receive many requests for data and so, as their time is precious, become unwilling or unable to provide data. A good principle is to present the request for data neatly and politely, ensuring that the data can be provided easily, that they are not too sensitive, that the study is of interest to the respondent and that the respondent will get a return commensurate with the effort expended to provide the data. At least, a summary of the research report should be offered and then *provided*.

Having identified the data required and the source(s) of the data (often in the form of the sample), commonly by organisations, the next step is to identify the most appropriate respondent in each organisation. For a study of quality, the job title to look for could be 'quality manager', 'director of quality', 'quality controller'. The real issue is to determine which person is at the appropriate level in the organisation to be able to provide the data required for the research. An examination of the organisation's website and/or an initial telephone call will be useful to determine who, if anyone, is the appropriate person – preferably by name as well as job title. If the person identified can be contacted by telephone or by email, the study can be explained *briefly* and, it is hoped, their agreement to provide data obtained. The respondent should be advised of the nature and extent of data required, including the time required for completion of any questionnaire or interview. The time needed should be obtained from the piloting, so that the respondent can understand their commitment. Ensure that the time allowed is reasonably accurate.

Commonly, anonymity will not be necessary, although confidentiality may be advisable, in order to obtain fuller and more readily given responses. The assurances can be given verbally but should be confirmed in writing in the formal letter of request for response in which the purpose and legitimacy of the research should be explained. It is useful if the letter contains an explanation of the research, the envisaged outcomes, benefits and purpose of the work as well as an explanation of its role in a degree course etc.

Despite assurances of confidentiality, such as, '... any data provided will be treated as confidential and used for the purposes of this research only; the identity of respondents will not be revealed', respondents may require further restrictions to apply concerning publication of results. Such restrictions should be considered carefully as they could 'stifle' the work and its value. It is legitimate, of course, to protect trade secrets, but 'vetting' of the contents of research reports by 'commercial' organisations solely because they have provided data should be avoided. However, for 'social investigations', including those employing ethnography, it is advisable to check the results and their interpretation with a 'focal respondent' to ensure that the meanings attributed to the data and the results of analyses are appropriate because the research is intended to operate and determine meaning 'through the lens' of the respondents.

'Vetting' should be avoided as it is likely to restrict the report, by suppressing or removing sections etc. and it may introduce bias. Trade secrets etc. can be protected by confidentiality and/or anonymity measures, describing respondents as 'companies A, B and C', and/or by restricting publication of the results etc. for a period of time or by limiting their scope of circulation. If confidentiality is to be provided, ensure that it really is so – care in the report can be destroyed by polite naming of respondents in 'Acknowledgements'.

As case study type data collections are in-depth, they are more likely to encounter commercially sensitive issues. Hence, extra care may be necessary to ensure confidentiality. As it is likely to be more problematic to obtain the data, so cultivation of contacts can pay dividends. Sections of the research, and report, may have to be treated differently where the issues tackled vary in sensitivity and, hence, in their confidentiality requirements.

Irrespective of the approach to data collection, piloting of the collection is vital. Executed well, amongst helpful, informed and appropriate respondents, it will reveal flaws in the data collection method and parameters (such as time required for the collection) can be determined. Modification and re-piloting will pay dividends by enhancing the rate of responses and quality of data obtained for analysis. Questions must be clear and precise and *not superfluous*. Remember, piloting should be carried out in full – to ensure the data can be analysed suitably and that the results of the analysis will address the research questions – hypothesis and objectives; proper piloting is more than testing the data collection instrument(s).

Example

If you send a questionnaire by post to Ms A.Y. Lai, City Construction, Beijing, it is rather silly to have:

Question 1: name of respondent
Question 2: organisation's name and address

as they are obviously known and their inclusion will have the likely effect of upsetting the prospective respondent by indicating a lack of sense, sensitivity, attention to detail and courtesy by the researcher. 'How many other questions are unnecessary?' might be asked.

Questions must be clear and easy to answer.

> *Example*
> Instead of asking the following questions:
>
> - Do you prefer tea or coffee?
> - How many people did your organisation employ in 2001?
>
> It would be preferable to ask:
>
> (1) Please tick which of the following beverages you prefer:
>
>   (a) tea   ☐
>   (b) coffee   ☐
>   (c) no preference ☐
>
> (2) Please tick the appropriate box to indicate how many people your organisation employed (on average) during calendar year 2001.
>
> | 0–9 | 10–39 | 40–99 | 100–199 | 200–499 | 500–999 | 1000+ |
> | --- | --- | --- | --- | --- | --- | --- |
> | ☐ | ☐ | ☐ | ☐ | ☐ | ☐ | ☐ |

In the original version of question (1), unless the respondent replied 'tea', 'coffee' or 'don't know', the answer would be ambiguous – consider what 'yes' might indicate. The original question (2) suggests that an answer to the nearest one employee is required, which is unlikely to be the case; the revised response indicator notes the level of accuracy appropriate. In a large organisation, the exact number of employees may vary from day to day, it may not be known exactly by the respondent and it might be difficult to discover precisely. Hence, a practical level of detail appropriate for the research and readily available to the respondents is required. All questions must be material to the research, not just of passing interest. The '20-question' guide is often quite useful and about the right number. Clearly, certain, usually qualitative, research projects must collect all possible data in order to search for patterns etc., but in a constrained or quantitative research project, such an approach is unlikely to be employed.

The more broad a concept is, the greater is the likelihood that it encompasses underlying dimensions, each of which reflects a different, important aspect of the overall concept. Lazarsfeld (1958) suggested a four-stage process for determining data requirements:

- Imagery
- Concept specification (dimensions)
- Selection of indicators
- Formation of scales/indices (for measurements).

Theory and literature may be helpful in identifying the dimensions and indicators; *a priori* establishment of the dimensions is very useful in forming indicators of the concept. Factor analysis is useful in examining the relationships between the indicators and how they relate to the concept.

Owing to problems with collecting original data, researchers increasingly 'pool' data – data are collected by groups of researchers who wish to investigate different but related topics – a common data set can represent an efficient compromise. Data collected by others can be used provided full details of the data collection methods, samples etc. are available as well as the data themselves. Published data are a good source and should be used as much as possible, not only for savings of time and cost, but also due to availability and convenience – such data can be used for a preliminary or background study – perhaps to provide the context for a detailed study involving collection of specific, original data.

If the particular data desired cannot be obtained, it is likely that surrogate measurements can be made. If treated with care, such measurements can yield very good approximations to what was sought originally, but do note the differences in the measurements and any changes in assumptions necessary.

> *Example*
> In a research project designed to investigate the costs incurred by contractors in their tendering for projects, it is unlikely that it will be possible to collect data from contractors of their costs of tendering – due to commercial sensitivity and non-recording of the data by the firms. Surrogate data in the form of resources used in tendering for a hypothetical project(s) may be collected to which 'nominal', standard costs could be applied (the costs being derived from job advertisements etc.) to yield a reasonably appropriate cost level.

Therefore, data form the essentials of a research project; after collection and analysis, their interpretations must begin to yield meaning in the context of theory and literature. For postal, email or web-based questionnaires, a date for commencement of analysis should be set to allow adequate time to obtain responses, including issuing polite reminders. A date for receipt of responses by the researcher should be specified clearly on the questionnaire and in a covering letter, usually 2–3 weeks from respondents' receipt of the questionnaire; at that time, non-respondents can be given reminders by post, telephone, fax or email. However, if only non-respondents are targeted, this means that the researcher has a referencing mechanism and so, responses are not anonymous. The date for commencement of analysis should be adhered to, and any responses received after that date excluded.

Especially, if the topic under investigation changes rapidly (as in IT) or is susceptible to 'shocks' (such as effects of political decisions/policy changes), it is important that the period for obtaining responses is both sufficiently long to secure sufficient responses but is also quite short to minimise the possible effects of changes. Of course, any changes which are found to occur during the response period must be noted and taken into account in the analyses, production of results and conclusions drawn. Any changes mean that the early and late responses may be incompatible (Clausen & Ford 1947).

Moser and Kalton (1971) note six primary conditions for (postal) questionnaires, apart from non-responses:

- '... the questions ... [must be] ... sufficiently simple and straightforward to be understood with the help of printed instructions and definitions ....
- ... the answers ... have to be accepted as final .... There is no opportunity to probe ....
- ... are inappropriate where spontaneous answers are wanted ....
- ... the respondent ... can see all the questions before answering any of them, and the different answers cannot therefore be treated as independent.
- ... [the researcher] ... cannot be sure that the right person completes the questionnaire.
- ... there is no opportunity to supplement the respondent's answers by observational data.'

They note that 'Ambiguity, vagueness, technical expressions and so forth must be avoided ... A mail questionnaire is most suited to surveys whose purpose is clear enough to be explained in a few paragraphs of print ... and in which the questions require straightforward and brief answers.'

As in any survey, non-responses present a problem, not just because they reduce the size of the sample which can be analysed but, more notably, because they may represent a body of opinion which, although unknown, may be significantly different from that which has been expressed by those who did respond. It may be the case also that the responses given by those for whom a follow-up reminder was necessary to obtain their responses form another group with a 'cluster' of opinions; however, analysis can reveal such clusters. Hence, the need for keeping good records of how, and when, responses were received.

However non-responses are dealt with – initial follow-up reminders, re-surveys etc. – note that, 'Non-response is a problem because of the likelihood – repeatedly confirmed in practice – that people who do not return questionnaires differ from those who do ... It has also been shown frequently that response level is correlated with interest in the subject of the survey ... If the response rate is not high enough to eliminate the possibility of serious bias, care must be taken to judge the extent of the unrepresentativeness and to take account of it in making the final estimates' (Moser & Kalton 1971). Further, they note a tendency for there to be upward bias in terms of both educational and socio-economic level of those who do respond to general mail questionnaire surveys. That may be reinforced by the personalities/personal characteristics of the people sampled – those who tend to be prompt, conscientious, etc. are more likely to respond and to do so early.

If it is not possible to investigate non-respondents through a 'follow-up study' of a sample of non-respondents (of the majority category amongst early respondents, late respondents and non-respondents), Clausen and Ford (1947) suggest that late respondents' responses may approximate to the views held by non-respondents or that trends in responses received over time can be used to indicate non-respondents' perspectives.

Although some researchers (e.g. Franzen & Lazarsfield 1945) suggest that if the population, and, therefore, the sample, is homogeneous in respect of the topic under investigation, then the problems of non-response (and early/late response) bias are not material. However, it is unlikely that homogeneity of the population regarding the topic would be known *a priori*; even if 'technical' homogeneity were apparent, interest and other sources of potential bias remain likely.

Clearly, there is no single best way of dealing with non-responses. Thus, response levels must be noted, along with the results obtained from the responses. They should be grouped as appropriate, as well as overall. Discussion of the impact of non-responses should be noted clearly as such. Separate what has been found out via the responses from what can be postulated reasonably to allow for non-responses.

For interview surveys, the issue of non-responses can be dealt with as the sample of respondents is being assembled. Interviews may be used to obtain greater 'depth' following a postal questionnaire survey or/and to obtain information about non-responses to the questionnaire. However, interviews may be subject to various sources and types of error and bias. Some have been noted earlier; those of the interviewer wishing to obtain support, those of interviewees wishing to 'please' the interviewer etc. It is important to remember that interviewing is a human, social process, however it is executed and, as such, it is subject to the interactions which occur between the participants. If the participants like each other, the process will be different and the responses may be different from a situation in which the participants dislike each other, even if it is only a first impression. The subject matter can influence such personal interaction too.

Thus, as well as training interviewers to ask questions and probe objectively, they must be able to record the situation of the interview accurately as well as the interviewees' responses. Cannell and Kahn, in Lindzey and Aronson (1968), note three conditions necessary for successful interviews:

- *accessibility*, to the interviewee of the information required;
- *cognition*, the interviewee's understanding of what is required;
- *motivation*, of the interviewee to answer the questions accurately.

Atkinson (1967) distinguishes three types of questions:

- factual
- knowledge
- opinion.

Opinion questions are the most sensitive category; factual are least sensitive. The more sensitive the category of questions, the more important it is that the questions are not perceived by the respondent to be 'threatening'. A 'threatening' question reduces the response rate to individual questions, if not to the entire interview or questionnaire.

Kahn and Cannell (1957) note five primary reasons for inadequate responses:

- partial responses
- non-responses
- irrelevant responses
- inaccurate responses
- verbalised response problem, where the respondent gives a 'reason' for not answering.

Many of the possible causes of inadequate or non-responses (including 'threatening' questions) should be removed by good piloting. Piloting should ensure also that the time required of interviewees is not unreasonable and that probing is being employed appropriately. Further, it will indicate whether the method of recording both responses and interview 'situations' etc. is adequate.

A particular feature of piloting is that it should be followed through with initial analysis, production of results etc. so that the data provision and use are checked thoroughly.

Bryman and Cramer (1994, p. 64) note that, 'if a question is misunderstood by a respondent when only one question is asked, that respondent will not be appropriately classified; if a few questions are asked, a misunderstood question can be offset by those which are properly understood'.

Any form of data collection based on self-reporting may be subject to bias through the respondents yielding skewed responses as being socially desirable etc. The positive approach attempts to solve such problems – respondents are required to choose between two alternatives or to rank a list of items, supplemented by rating.

- Ranking: → hierarchy
- Rating: → degrees of importance.

Thus, interpretation can focus on major aspects (high rating as well as high ranking).

Rating is often omitted, even though it is more useful than ranking as it is more revealing in determining overall significance than ranking alone (note also, that rankings can be produced from ratings), where items may already have been determined to be important via theory or past research. In such cases, the researcher must consider the applicability or transferability of that theory and past research.

Two further issues which concern data are:

- validity
- reliability.

Validity is discussed in Chapter 5.

Reliability concerns the consistency of a measure. *External reliability* concerns the consistency of a measure over time – often measured by re-test reliability – ensuring that the tests are sufficiently far apart in time that the respondents do not answer the second test by recalling their responses to the first. Any events which have occurred between the occasions of the tests should be taken into account in examining the results of the test and retest.

The concept of statistical sampling concerns obtaining a sample which is adequate to represent the population under investigation with sufficient confidence. Even where non-statistical sampling methods are employed, such as a single case study of a construction project using ethnomethodology to investigate relationships, there is likely to be a desire to consider any validity of the results beyond the sample obtained.

Strictly, of course, unless the research is designed to obtain data and, hence, results, which are valid beyond the confines of the data or sample, the findings are valid for the subject data only. The issues concern statistical inference, deduction and induction. No matter how well a research is designed, it remains subject to the data actually obtained as responses, experimental measurements etc. in determining its validity, both internal and external. Hence, following the discussions of sampling, responses and measurement, the issue of adequacy of data response remains.

Clearly, data collection should be statistically designed to provide sufficient expected responses, so that the desired validity will be satisfied through sample selection and sizing, allowing for anticipated response rates. For example, if 'small number' statistics are to be avoided, a minimum of 32 usable responses is necessary; given an expected response rate of 30% for a postal questionnaire, a sample size of over a hundred, minimum 107, is necessary.

Especially for surveys (and notoriously, the enormously popular postal questionnaires) it is important to consider response rates. As well as numbers of responses obtained, in order to evaluate the validity of the data, one must consider the likelihood of non-responses from various groups being different, due to the issues of non-responses likelihood of being different. Indeed, responses which are obtained after 'follow-up' requests may constitute an intermediate group between immediate responses and non-responses.

Fowler (1984) noted 'there is no agreed-upon standard for a minimum acceptable response rate'. There are a variety of perspectives to evaluate in determining the responses to be sought; these relate to the population, the sample, the tests and, most especially, the validity and applicability of the findings. Rigorous calculations and judgements are required.

Summary

This chapter has addressed a variety of issues concerning collection of data. In particular, the issues of sampling have been examined – size and structure. The various types of data have been discussed and the tests to which they may be subjected. Approaches to elicit data from respondents have been outlined along with the necessity to preserve confidentiality and anonymity in some instances. The issue of response rates is important and must be taken into account when deciding the size of sample. Piloting is vital to ensure data provision by respondents is easy and the requirements are clear.

References

Ahmad, I. & Minkarah, I. (1988) Questionnaire survey on bidding in construction, *ASCE Journal of Management in Engineering Divisions*, **4**(3), 229–243.

Atkinson, J. (1967) *A Handbook for Interviewers: a Manual for Government Social Survey Interviewing Staff, Describing Practice and Procedures on Structured Interviewing.* Government Social Survey No. M136, HMSO, London.

Baker, M. (1997) Membership categorisation and interview accounts. In: *Qualitative Research: Theory, Method and Practice* (ed. D. Silverman), pp. 130–143, Sage, London.

Bell, J. (1993) *Doing Your Research Project*, Open University Press, Milton Keynes.

Bryman, A. & Cramer, D. (1994) *Quantitative Data Analysis for Social Scientists*, revised edn, pp. 62–70, Routledge, London.

Cattell, R.B. (1966) The scree test for the number of factors, *Multivariate Behavioural Research*, **1**, 140–161.

Clausen, J.A. & Ford, R.N. (1947) Controlling bias in mail questionnaires, *Journal of the American Statistical Association*, **42**(240), 497–511.

Cleland, D.I. & King, W.R. (1983) *Systems Analysis and Project Management*, 3rd edn, McGraw-Hill, Singapore.

Cochran, W.G. (1977) *Sampling Techniques*, 3rd edn, John Wiley, New York.

Cook, T.D. & Campbell, D.T. (1976) The design and conduct of quasi-experiments and true experiments in field settings. In: *Handbook of Industrial and Organizational Psychology* (ed. M.D. Dunette), Rand McNally, Chicago.

Cook, T.D. & Campbell, D.T. (1979) *Quasi-experimentation: Design and Analysis Issues for Field Settings*, Rand McNally, Chicago.

Drenth, P.J.D. (1998) Research in work and organizational psychology: principles and methods. In: *Handbook of Work and Organizational Psychology*, Vol. 1, 2nd edn, (eds P.J.D. Drenth, H. Thierry & C.J. de Wolff), pp. 11–46, Psychology Press, Hove.

Drenth, P.J.D., Thierry, H. & de Wolff, C.J. (eds) (1998) *Handbook of Work and Organizational Psychology*, 2nd edn, Vol. 1, Psychology Press, Hove.

Eisenhardt, K.M. (1989) Building theories from case study research, *Academy of Management Review*, **14**(4), 532–550.

Fiedler, F.E. (1967) *A Theory of Leadership Effectiveness*, McGraw-Hill, New York.

Fowler, F.J. (1984) *Survey Research Models*, Sage, London.

Franzen, R. & Lazarsfield, P.F. (1945) Mail questionnaires as a research problem, *Journal of Psychology*, **20**, 293–320.

Garfinkel, H. (1967) *Studies in Ethnomethodology*, Prentice-Hall, Englewood Cliffs, NJ.

Holstein, J.A. & Gubrium, J.F. (1997) Active interviewing. In: *Qualitative Research: Theory, Method and Practice* (ed. D. Silverman), pp. 113–129, Sage, London.

Jick, T.D. (1979) Mixing qualitative and quantitative methods: triangulation in action, *Administrative Science Quarterly*, **21**(4), 602–611.

Kahn, R.L. & Cannell, C.F. (1957) *The Dynamics of Interviewing; Theory, Technique and Cases*, Wiley, New York.

Krejcie, R.V. & Morgan, D.W. (1970) Determining sample size for research activities, *Educational and Psychological Measurement*, **30**(3), 607–610.

Lazarsfeld, P.F. (1958) Evidence and inference in social research, *Daedalus*, **87**, 99–130.

Lindzey, E. & Aronson, G. (1968) *The Handbook of Social Psychology, Vol. 2, Research Methods*, 2nd edn, Addison-Wesley, Reading, MA.

Liu, A.M.M. (1997) A model for application of Conjoint Analysis in value management. In: *Procurement – a Key to Innovation* (eds C.H. Davidson, A.A.M. Tarek), pp. 413–422, CIB Proceeding, CIB Publication no. 203, CIB.

Liu, A.M.M., Shen, L. & Lam, B. (1994) Conjoint Analysis in Contractors' Project Selection Decisions Proceedings of the ARCOM 10th Annual Conference. Loughborough University, UK, September 1994 (eds R.M. Skitmore and M. Betts), University of Salford, 1994, Vol. 2, pp. 640–649.

MacRae, A.W. (1988) Measurement scales and statistics: what can significance tests tell us about the world? *British Journal of Psychology*, **79**, 161–171.

Malhotra, N.K. & Birks, D.F. (1999) *Marketing Research. An Applied Approach*, Prentice-Hall, London.

Miller, J. & Glassner, B. (1997) The 'inside' and the 'outside': finding realities in interviews. In: *Qualitative Research: Theory, Method and Practice* (ed. D. Silverman), pp. 99–112, Sage, London.

Morgan, D.J. (1997) *Focus Groups as Qualitative Research*, 2nd edn, Sage, London.

Moser, C.A. & Kalton, G. (1971) *Survey Methods in Social Investigation*, 2nd edn, Dartmouth, Aldershot.

Newcombe, R., Langford, D.A. & Fellows, R.F. (1990) *Construction Management 1: Organisation Systems*, Mitchell, London.

Osgood, C.E., Suci, G.J. & Tannenbaum, P.H. (1957) *The Measurement of Meaning*, University of Illinois Press, IL.

Pettigrew, A.M. (1997) What is processual analysis?, *Scandinavian Journal of Management*, **13**(4), 337–348.

Rogers, E.M. & Kincaid, D.L. (1981) *Communication Networks: Towards a New Paradigm for Research*, Free Press, London.

Rowlinson, S.M., Ho, T.K.K. & Po-Hung, Y. (1993) Leadership style of construction managers in Hong Kong, *Construction Management and Economics*, **11**(6), 455–465.

Shash, A.A. (1993) Factors considered in tendering decisions by top UK contractors, *Construction Management and Economics*, **11**(2), 111–118.

Yeomans, K.A. (1968) *Statistics for the Social Scientist 2: Applied Statistics*, Penguin, Harmondsworth.

Chapter 7

Data Analysis

The objectives of this chapter are to:

- introduce logical procedures for *analysing data*;

- demonstrate the value of *plotting data* in various ways;

- discuss some primary statistical methods for analysing data – non-parametric tests and parametric tests;

- introduce *other analytical techniques* that are applicable in management and construction research.

Analysing data

It has been noted that the choice of data collected should be determined by the outputs required from the research, given constraints of practicality. One essential consideration, within the systems perspective of Chapter 6, is the analysis of the data which is to be undertaken. Unfortunately, it has become all too common, especially for new and enthusiastic researchers, to plunge into the most complex statistical techniques they can find, often using computer packages in a 'black box' fashion, only to 'emerge' sometime later rather bemused.

The preferable approach is to consider, evaluate and plan the analysis in a similar way to planning the whole research project. Geddes (1968), the 'father of town planning', advocated the method of:

- survey
- analyse
- plan.

A sensible way to ensure that the methods selected are appropriate.

Not all research projects yield data which are suitable for statistical analyses, and even those which do may require only simple manipulations of small sets of data. Computing helps but is not essential – it makes calculations quicker and easier – an advantage which can cause major problems too. Problems can occur as packages are, increasingly, 'user friendly', some to the extent that the researcher may not be sufficiently aware of the statistical bases and assumptions of the tests. In the early days of the particular package called Statistical Package for the Social Sciences (SPSS), an expert user obtained only a little printout, as the appropriate tests had been specified and performed on the data; a novice user received a 'mountain' of printout as everything available had been performed on the data – the user had not specified the appropriate tests and so, the package had executed all tests available and so, the user was left with the problem of sorting through the printout to extract useful elements. So, 'user friendly' programs are very helpful, but can cause the 'black box' syndrome; it is important to remember that, to be useful, tests must be valid and understood.

No matter what is the nature of the data collected, it is appropriate to begin analysis by examining the raw data, using scatterplots, to search for patterns. Of course, a pattern or a relationship may be expected from the review of theory and literature – one may have been hypothesised. Alternatively, for fundamental studies where theory and literature do not exist to any great degree, the search for patterns and relationships in the data and the identification of major variables may constitute the total analysis for the research project.

For data sets in topics which have an extensive body of theory and literature, it is good practice to search the data, *with an open mind*, for themes and categories. Of course, usually, such scrutiny will serve to confirm the themes and categories etc. found in the theory and literature, but the researcher must be prepared to discover differences in the data from what theory and previous findings suggest will occur. Societies are dynamic, so changes over time should be expected and methods of data collection and analysis must be sufficiently rigorous to detect them.

Qualitative data can be difficult and laborious to analyse – they must be handled systematically; a requirement which is easier with quantitative data. Categorisation of qualitative data may rely on the researcher's opinion (objectivity based on rationality, expertise and experience); it may be useful to construct a set of guidelines initially, and to confirm or amend and supplement them on a 'first pass' of the data. A 'second pass', using the completed categorisation, will ensure that all of the data, especially the data considered early in the first pass, are categorised consistently. For large sets of data, a 'piloting' exercise using a sample may serve as the first pass. Essentially, the approach arises as part of grounded theory. In such an exercise, it is necessary to consider each transcript so that the contexts of words are not lost. For example, the word 'tap' has a variety of meanings in England; equally, slang terms vary in meaning.

Thus, for much 'human-oriented' research in construction, the objective is to analyse peoples' behaviours, including their causes and consequences, as manifested in actions and symbols – notably oral and written language. Some data may be documented outside of the research, such as 'newspaper' articles, others may be obtained specifically (e.g. questionnaires), and others may be collected through researcher observation, during interviews, notes of meetings, notes of 'shadowings' etc. Respondents' diaries may be obtained for analysis although, to avoid bias, it is preferable to obtain diaries which have been kept by respondents as a matter of course or daily routine in their activities. Where data are in an oral form, such as tape recordings of interviews, usually, it is best to transcribe the data prior to analysis, in order to aid clarity of data and consistency of analysis. For more detailed analyses, transcripts should include notes of pauses, interruptions, hesitations etc. in the responses; indeed, video recording may be necessary to collect hand gestures, facial expressions and other forms of non-verbal communication all of which contribute to the communication of meanings by the respondents.

In most contexts, visual aids and diagrams can be extremely helpful in analysing data, as patterns and relationships often emerge. Such diagrams should comprise (as near as is practical) the raw data; this is relatively simple for quantitative data but will be the result of the initial scrutinies where categories of qualitative data are required. For 'second hand' data, such as statistics published by government, the raw data may not be available, or they are likely to be inconvenient to access, so the published data must be used.

Although the patterning revealed by examination of quantitative data may be quite straightforward, two considerations involving potential difficulties may occur. The first is that theory and literature, by advancing relationships between variables via hypotheses, may lead to other possible relationships not being considered. The second is to ensure that, especially if using a computer, the data have been input correctly. Usually, data have to be coded. Whilst this tends to be simple and obvious for quantitative data, coding may distort, or be part of the analysis, of qualitative data. Ensure the coding is both easy to use and understand and is of an appropriate level of detail. Too little detail will yield 'conglomerate categories' which do not reveal meaning. Too much detail will not only produce allocation problems, but also yield so many categories that there may be almost a category for each item of data, rendering analysis unwieldy.

Moser and Kalton (1971) note that the sets of data from each respondent should be subjected to an editing process before coding. The editing should check the data sets for completeness – so that any gaps may be filled, if possible; for accuracy – to check or verify any apparent inconsistencies; for uniformity – so that responses are in the same form for coding, notably where interviews have been employed or/and semantic responses may be used to produce a frame or keyword set of response contents: an elementary form of content analysis.

As noted above, many analyses of qualitative data concern searching the data for patterns of various types, so that hypothesised relationships can be established for subsequent investigation and testing – perhaps, by more quantitative methods. In seeking patterns, two main approaches may be employed, either individually or together, to search the data for patterns (e.g. as in grounded theory), or to employ theory and literature to suggest likely 'rational' patterns; however, the latter approach, if adopted alone, may result in the researcher's missing new, and potentially important, relationships in the data – an 'open' mind, as free as possible from preconceptions, is likely to be most appropriate and revealing.

Many qualitative approaches are not subject to particular analytic techniques with prescribed tests, as is common in quantitative analyses, such as conversation and discourse analyses. Instead, they involve scrutiny of transcribed texts of discussions, statements etc. so that not only is the content analysed but also the linguistic context is considered, in order to establish the meanings, intentions, interpretations etc. of the people concerned. Hence, the researcher must develop sensitivity to the people, their language and the way in which language may be used. Discourse analyses involve many readings of the texts being analysed, so that iterative formulation, testing and revision of hypotheses concerning the discourses in the texts may occur. For this reason, the context of the texts of the discourse is important as an indicator of possible or likely purpose.

Content analysis may be employed, at its most simplistic, to determine the main facets of a set of data, by simply counting the number of times an activity occurs, a topic is mentioned etc. However, even for such apparently straightforward analysis, awareness and interpretations by the researcher are likely to be necessary – a number of actions, although different, may be very similar in purpose; several words or phrases etc. may have very similar meanings – and so, the boundaries of categories must be established to a sufficient, but not overwhelming, extent for the analysis. For many content analyses, it is important to have a sound theoretical basis to assist development and testing of hypotheses – such as non-verbal behaviours of people in meetings. This is so that actions, such as those which indicate aggression, nervousness etc., in that society can be identified. Clearly, virtually identical behaviours can have different meanings depending on the contexts so, here too, the situation should be considered holistically for analysis. Thus, once the categories of data have been established, a content analysis will yield quantitative data for each content category. Some analyses may be required to yield such descriptive results only; others may wish to continue to investigate relationships – using correlations or more multidimensional analyses.

The initial step in content analysis is to identify the material to be analysed. The next step is to determine the form of content analysis to be employed – qualitative, quantitative or structural; the choice is dependent on, if not determined by, the nature of the research project. The choice of categories also will depend upon the issues to be addressed in the research, if they are known.

In qualitative content analysis, emphasis is on determining the meaning of the data. Data are given coded allocations to categories and groups of 'respondents' from whom the data were obtained are fitted to these categories, so that a matrix of categorised data against groups is obtained. Statements etc. can be selected from each cell of the matrix to illustrate the contents of each of the cells. As in any allocation mechanism, the categories should be *exclusive*, i.e. data assigned to one category only, and *exhaustive*, i.e. categories cover the research topic comprehensively.

Quantitative content analysis extends the approach of the qualitative form to yield numerical values of the categorised data – ratings, frequencies, rankings etc., which may be subjected to statistical analyses. Comparisons may be made and hierarchies of categories may be examined.

Structural content analysis concerns determination and examination of relationships between categories of data and between groups where this is appropriate. The rules used to determine relationships will depend on the aims of the research project.

Irrespective of the form of content analysis employed, there may be a tendency to consider only what is mentioned in the transcript or data; in some cases, what is omitted is of great importance and astute researchers should consider such omissions. Further, not just categories of data but combinations of categories may be important too.

The structures of qualitative data may be investigated using Multidimensional Scalogram Analysis – MSA (Lingoes 1968). MSA facilitates setting research subjects (people in a survey or respondents) and the variables under consideration, to be shown on one diagram. Given the research subjects and the dimensions of the variables, a matrix can be constructed. By grouping subjects' scores against the variables, the numbers in the cells of the matrix can be ascertained. Scores for the variables are best kept simple (say in the range 1–4). Diagrams are produced which represent the results to aid comparisons between subjects. This is discussed and demonstrated by Wilson (1995).

To a large extent, the nature of diary data depends on what, if any, structuring of the diary is imposed; especially for 'freely composed' diaries, various analytic techniques are available but, initially, a content analysis is likely to be helpful. Once the contents of the diaries have been categorised, and, hence, given some common structure, it may be appropriate to proceed with more quantitative analyses, as well as qualitative and descriptive ones.

The purpose of analysing the data is to provide information about variables and, usually, relationships between them. Hence, as research in a topic becomes more extensive, quantitative studies may be undertaken to yield statistical evidence of relationships and their strengths; statistics are useful in determining directions of relationships (*causalities*) when combined with theory and literature.

However, the purpose of analysis is to provide evidence of relationships and to aid understanding; in a context of management, it is to support decision making – hence, the importance of *inference*. Inference is what follows logically from the evidence, and it is important to know how valid those inferences are. Popper (1989, pp. 317–318) notes that, 'a rule of inference is valid if, and only if, it can never lead from true premises to a false conclusion'. A summary of some quite simple statistical techniques which are used extensively in analyses of research data follows. Most computer statistics packages can perform the numerical manipulations but *the researcher must understand what is being done*. Beware of the idiot machine!

Plotting data

Once the data have been collected, it is helpful to produce a diagram or graph of those data (a 'scatter plot' of the raw data). Such plots will help to indicate the nature of distribution of the data and relationships between them such that appropriate statistical techniques, if any, may be employed in analysis. Not all data lend themselves to plotting in the form of a graph. For dichotomous variables, such as 'yes' or 'no' responses, cross-tabulations, or contingency tables, are used to detect patterns. The next step is to undertake a statistical analysis, the most usual of which is the χ^2 (chi-square) test.

A table of desired and actual responses, such as to questionnaires sent and received back, noting proportion usable for the research, is useful to demonstrate the sampling attempted and realised. Use both actual numbers and percentages to convey maximum information. Having depicted the data (responses) being considered, focus can fall on analyses. Analysis examines responses so that patterns and relationships between variables can be discovered and quantified, *with theory helping to explain causation*. Consider, e.g. percentage changes in costs observed in a sample of projects. The data are summarised in Fig. 7.1.

% Change	−10 to −5.1	−5 to −0.1	0 to 4.9	5 to 9.9	10 to 14.9	15 to 19.9	20 to 24.9	25+
No. projects	0	3	10	21	14	6	1	0

Fig. 7.1   Frequency distribution ($n = 55$).

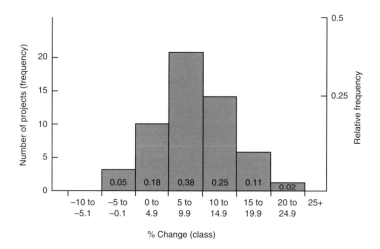

Fig. 7.2   Bar diagram (relative frequencies may not sum to 1.0 due to rounding).

Now consider the diagrammatic representation shown in Figs. 7.2 and 7.3 of the data in Fig. 7.1.

Histograms have the particular property that the area of each rectangle represents the proportion of the number of observations in that class; this property does *not* apply to a *bar diagram*.

It is quite simple to convert a frequency polygon into a *frequency curve*. A frequency curve gives an indication of the shape of the sample distribution, and, consequently, of the population distribution, provided good sampling techniques have been used.

An *ogive* (Fig. 7.4) is a form of cumulative frequency distribution curve. Using the upper boundary of the class intervals yields the 'more than' cumulative curve.

Presentation of 'raw' data provides the greatest detail. However, even if ordered in some way, the data may not be easy to interpret. In presenting data, it is common for detail to be sacrificed, so that intelligibility is improved by the use of tables and diagrams. In any event, it is helpful if the 'raw' data are presented in an appendix to the research report; data are valuable for further

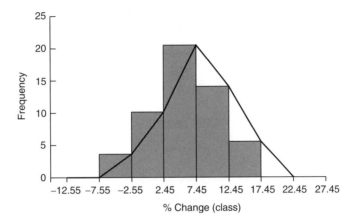

Fig. 7.3   Frequency polygon or *histogram*.

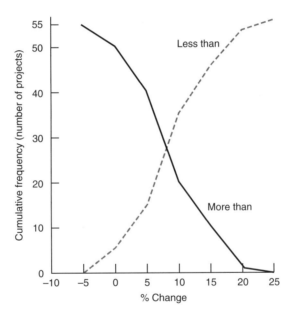

Fig. 7.4   'More than' and 'less than' ogives.

studies, tests etc. as well as for checking and verification of the instant analysis and results.

The statistical methods noted above are purely *descriptive* – as measures of the data obtained. They do not, of themselves, constitute analysis.

For research, it is important that the data, however obtained, are subjected to appropriate and rigorous analysis to assist determination of meaning. Some common methods of statistical analysis are considered below.

Note that, in many instances, the (numerical) results of statistical tests provide only partial information – the critical aspect is whether the result of the test, given the sample size etc., is *statistically significant* and, if so, at what level of confidence/significance. It is also essential to ensure that appropriate test methods have been used.

Statistical methods

Some common statistical methods used in data analysis are discussed in this chapter:

- non-parametric tests
  - sign test
  - rank-sum test
  - chi-square test
  - goodness of fit
- parametric tests
  - *t*-test
  - ANOVA (analysis of variance)
- regression and correlation
- time series
- index numbers.

Parametric and non-parametric tests are available for testing hypotheses related to differences in the metric and non-metric scales. In the parametric case, the *t*-test is used to examine hypotheses related to the population means. In the non-parametric case, popular tests include the chi-square, binomial test, and Mann–Whitney *U*-test (an example is given in Table 7.1).

Non-parametric tests

Non-parametric tests are distribution-free and so, are more flexible in application.

Sign test

The sign test examines paired data using positive (+) and negative (−) signs.

Table 7.1   Hypothesis testing.

Sample	Application	Level of Scaling	Common Data Analysis Method
One sample	Distributions	Non-metric	Kolmogorov–Smirnov (K–S) one sample test and chi-square goodness of fit
	Means	Metric	t-test; z-test
	Proportions	Metric	z-test
Two independent samples	Distributions	Non-metric	K–S two sample test
	Means	Metric	Two groups t-test; F-test (ANOVA)
	Proportions	Metric	z-test
		Non-metric	Chi-square test
	Rankings/medians	Non-metric	Mann–Whitney U-test

Source: Malhotra and Birks (1999).

Example   The sign test
A sample of architects are asked to rate the performance of two types of roofing tiles, A and B; the scorings are: excellent $= 5$, to very poor $= 1$, on a 5-point Likert scale.

Architect No.	1	2	3	4	5	6	7	8	9	10	11	12
(1) Type A	5	4	5	2	1	2	3	2	5	5	3	1
(2) Type B	3	4	2	1	3	3	4	5	3	4	1	5
Sign [(1) − (2)]	+	0	+	+	−	−	−	−	+	+	+	−

No. +ve (p)	6
No. −ve (q)	5
No. 0	1
Total	12

H_O: $p = 0.5$ (there is no difference between tiles A and B)
H_A: $p \neq 0.5$ (there is a difference between tiles A and B)

excluding zeros:

$n = 11$
$\bar{p} = 6/11$ (proportion of 'successes')
$\bar{q} = 5/11$ (proportion of 'failures')

for *'no difference'*, $p H_O = q H_O = 0.5$

Standard error of the proportion:

$$\sigma_{\bar{p}} = \sqrt{\frac{pq}{n}} = \sqrt{\frac{(0.5)(0.5)}{11}}$$

$$= \underline{\underline{0.151}}$$

As H_A: $p \neq 0.5$ (i.e. concerned with larger OR smaller), a 2-tailed test is required.

At 0.05 level of significance and as np and $nq \nless 5$, the normal distribution approximates to the binomial, the z-value for 0.475 (i.e. 0.5 minus $\frac{1}{2} \times 0.05$) of the area under one tail of the normal curve is 1.96, then:

$$p H_O + 1.96\sigma_{\bar{p}} = 0.5 + (1.96)(0.151)$$

and

$$p H_O + 1.96\sigma_{\bar{p}} = 0.5 - (1.96)(0.151)$$

So, the range of acceptance is:

$$0.204 \rightarrow 0.796$$

The sample proportion,

$$\bar{p}\left(= \frac{6}{11} \right) = 0.545$$

As $0.204 < 0.545 < 0.769$, there is *no difference* in the architects' perceptions of the tiles (H_O is accepted).

Rank-sum tests

Rank-sum tests are used to test whether independent samples have been drawn from the same population.

The *Mann–Whitney U-test* is used when there are two samples, and the *Kruskal–Wallis K-test* is used when there are three samples or more.

Example   Mann–Whitney U

Consider rents of a particular type of building in two locations X and Y; the locations are quite close to each other; rents are expressed in $ per m^2 of floor area.

	Rents in Location X ($/m^2)	Rents in Location Y ($/m^2)
	30	39
	32	42
	40	54
	50	52
	55	36
	47	46
	48	45
	38	37
	41	33
	53	51

Rank	Rent	Location	Rank	Rent	Location
1	30	X	11	45	Y
2	32	X	12	46	Y
3	33	Y	13	47	X
4	36	Y	14	48	X
5	37	Y	15	50	X
6	38	X	16	51	Y
7	39	Y	17	52	Y
8	40	X	18	53	X
9	41	X	19	54	Y
10	42	Y	20	55	X

n_1 = number of buildings in sample 1
n_2 = number of buildings in sample 2
R_1 = sum of ranks in sample 1 (in X locations)
R_2 = sum of ranks in sample 2 (in Y locations)

From the example:

$n_1 = 10, n_2 = 10, R_1 = 106, R_2 = 104$
U = Mann–Whitney U-statistic

Using the standard form of the Mann–Whitney U-test:

The U-statistic is a measure of the difference between the ranked observations of the two samples:

$$U = n_1 n_2 + \frac{n_1(n_1 + 1)}{2} - R_1$$

$$= 10 \times 10 + \frac{10(11)}{2} - 106$$

$$= \underline{\underline{49}}$$

H$_O$: samples are from the same population
H$_A$: samples are from different populations

If H$_O$ applies, samples are from the same population and the U-statistic has a sampling distribution described by:

$$\mu_u = \frac{n_1 n_2}{2}$$

$$= \underline{\underline{50}}$$

μ_u = mean
z-value = confidence level required
α = level of confidence

$$\text{standard error, } \sigma_u = \sqrt{\frac{n_1 n_2(n_1 + n_2 + 1)}{12}}$$

$$= \sqrt{\frac{(10 \times 10)(21)}{12}}$$

$$= \underline{\underline{13.23}}$$

So H$_O$: $\mu_1 = \mu_2$
H$_A$: $\mu_1 \neq \mu_2$

$\alpha = 0.05$ (i.e. 95% confidence level)
z-value (using normal distribution) of $0.475 = \underline{1.96}$

$$\text{Limits: } \mu_u + 1.96\sigma_u = 50 + (1.96)(13.23)$$

$$= \underline{75.93}$$

$$\mu_u - 1.96\sigma_u = \underline{24.07}$$

As $24.07 < 49.0 < 75.93$, H$_O$ is accepted [i.e. $(\mu_u - 1.96\sigma_u) < u < (\mu_u + 1.96\sigma_u)$].

Note: If items of data have equal values, the rank assigned to each one is averaged.

Example   Kruskal–Wallis K

Tests have been carried out on three types of dumper trucks to determine the distance each travels on site using one gallon of fuel. The results, in miles, are:

Truck type A	6.0	6.8	5.7	5.2	6.5	6.1	
Truck type B	5.6	5.9	5.4	5.8	6.2	7.0	5.1
Truck type C	5.0	6.3	5.3	6.4	6.6	5.5	6.7

Rank	Distance	Type	Rank	Distance	Type
1	5.0	C	11	6.0	A
2	5.1	B	12	6.1	A
3	5.2	A	13	6.2	B
4	5.3	C	14	6.3	C
5	5.4	B	15	6.4	C
6	5.5	C	16	6.5	A
7	5.6	B	17	6.6	C
8	5.7	A	18	6.7	C
9	5.8	B	19	6.8	A
10	5.9	B	20	7.0	B

Using the standard form of the Kruskal–Wallis K-statistic:

$$K = \frac{12}{n(n+1)} \sum_{j=1}^{k} \frac{R_j^2}{n_j} - 3(n+1)$$

where: K = Kruskal–Wallis K-statistic
n_j = number of items in sample j
R_j = sum of the ranks of the items in sample j
k = number of samples
$n = n_1 + n_2 + \cdots + n_k$
= the total number of observations in all the samples

Type A	Rank	Type B	Rank	Type C	Rank
5.2	3	5.1	2	5.0	1
5.7	8	5.4	5	5.3	4
6.0	11	5.6	7	5.5	6
6.1	12	5.8	9	6.3	14
6.5	16	5.9	10	6.4	15
6.8	19	6.2	13	6.6	17
		7.0	20	6.7	18
	69		66		75

$$K = \frac{12}{20(20+1)}\left[\frac{(69)^2}{6} + \frac{(66)^2}{7} + \frac{(75)^2}{7}\right] - 3(20+1)$$

$$= 0.02857[793.5 + 622.3 + 803.6] - 63$$

$$= \underline{0.408}$$

According to Levin and Rubin (1990, p. 609), the K-statistic can be approximated by a chi-square distribution when all the sample sizes are at least 5. The number of degrees of freedom is $k - 1$.

$H_O: \mu_1 = \mu_2 = \mu_3$
$H_A: \mu_1, \mu_2, \mu_3$ are *not* equal
$\alpha = 0.05$

From tables of the chi-square distribution; with 2 degrees of freedom and 0.05 of the area on the right-hand tail, $\chi^2 = 5.991$.

As the calculated value of K is less than the tabulated value of χ^2, the sample lies within the acceptance region and so, H_O should be accepted; there is no difference between the trucks.

Chi-square (χ^2) test

The *chi-square test* is used to compare observed and expected frequencies of a variable which has three or more categories, to test whether more than two population proportions can be considered to be equal. Generally, the χ^2 distribution should *not* be used if any cell contains an expected frequency of less than 5.

Example
The numbers of male and female workers are noted over three construction sites. The researcher wishes to investigate possible sex discrimination between the construction sites, and so wishes to know if the data provide any evidence.

H_O: there is no difference in the proportion of female workers employed on each site.

	Site A	Site B	Site C	Total
Male	52	48	60	160
Female	13	15	12	40
Total	65	63	72	200

Normalising the data (rounding to whole numbers):

	Site A	Site B	Site C	Total
% Male	33	30	37	100
% Female	33	38	29	100
% Total	33	31	36	100

Hence, the question is, given a sample of 200 workers on the construction sites, is it reasonable for the female workers to be distributed A = 13; B = 15; C = 12, if there is no sexual discrimination between those sites?

Expected female workers (rounded):

$$\text{Site A: } \frac{65}{200} \times 40 = 13$$

$$\text{Site B: } \frac{63}{200} \times 40 = 13$$

$$\text{Site C: } \frac{72}{200} \times 40 = 14$$

Expected male workers:

$$\text{Site A: } \frac{65}{200} \times 160 = 52$$

$$\text{Site B: } \frac{63}{200} \times 160 = 50$$

$$\text{Site C: } \frac{72}{200} \times 160 = 58$$

$$\chi^2 = \sum \frac{(f_o - f_e)^2}{f_e}$$

where: f_o = observed frequency
f_e = expected frequency

so

$$\chi^2 = \left[\frac{(52-52)^2}{52} + \frac{(48-50)^2}{50} + \frac{(60-58)^2}{60} \right. $$
$$\left. + \frac{(13-13)^2}{13} + \frac{(15-13)^2}{13} + \frac{(12-14)^2}{14} \right]$$

$$= [0 + 0.08 + 0.067 + 0 + 0.308 + 0.286]$$

$$= \underline{\underline{0.741}}$$

The number of degrees of freedom of a χ^2 distribution is:

$$(\text{no. of rows} - 1)(\text{no. of columns} - 1)$$

Hence, in the example, the degrees of freedom are:

$$(2 - 1)(3 - 1) = 2$$

The tabulated value of χ^2 with 2 degrees of freedom and $\alpha = 0.5$ is 5.991. α = level of confidence.

Thus, as the calculated χ^2 is less than the tabulated value, the null hypothesis cannot be rejected; there appears to be no sexual discrimination in worker employment between the construction sites.

Goodness of fit

The goodness of fit of the data to a theoretical distribution is examined by the *Kolmogorov–Smirnov* test. The χ^2 test can be used for this purpose.

The Kolmogorov–Smirnov statistic, D_n, is the maximum value of the absolute deviation of $f_e - f_o$, where f_e and f_o are expected and observed relative cumulative frequencies. Critical values are tabulated such that if the calculated value of D_n is less than the tabulated value, the null hypothesis that the sample accords with the distribution postulated cannot be rejected and is, thus, accepted.

Parametric tests

Parametric tests assume that the distribution is known, or that the sample is large, so that a normal distribution (see Fig. 5.3, p. 139) may be assumed; equal interval or ratio scales should be used for measurements.

t-test

The *t*-test is used to determine if the mean of a sample is similar to the mean of the population.

$$t = \frac{\bar{x} - \mu}{\hat{\sigma}_{\bar{x}}}$$

where: $\hat{\sigma}_{\bar{x}}$ is the estimated standard error of the mean.
Degrees of freedom applicable are $(n - 1)$.
t_{tab} = value taken from tabulated *t*-distribution curve.

If $t_{calc} < t_{tab}$, the mean of the sample is not significantly different from the mean of the population at the selected level of significance (confidence). Usually, tests start by using the 95% level ($p = 0.05$), if the test at that level is passed, a higher level may be tested or *vice versa*. The 95% confidence level indicates that, although the data support the conclusion with 95% probability, there is a 5% chance that the conclusion is wrong.

The test may be used to examine the means of two samples:

$$t = \frac{\bar{x}_1 \sim \bar{x}_2}{\text{standard error of the difference in means}}$$

where: $\bar{x}_1 = \text{mean of sample 1}$
$\bar{x}_2 = \text{mean of sample 2}$
$\bar{x}_1 \sim \bar{x}_2 = \text{difference between the means.}$

Analysis of variance (ANOVA)

Analysis of variance (ANOVA) is based on a statistical method called the *F*-test, which is the ratio of the variance among conditions (between-groups variance) to the variance within conditions (within-groups, or error variance). The larger the between-groups variance relative to the within-groups variance, the larger the calculated value of F, and the more likely it is that the differences among the condition means reflect true effects of the independent variable rather than error variance.

ANOVA and *t*-test are developed in the context of experimental research to test the differences among the means of the experimental groups but they are also widely used to analyse data from non-experimental studies. In non-experimental studies, participants are not randomly assigned to groups (as in a true behavioural experiment) but are categorised into naturally occurring groups (e.g. architects, engineers etc.); then a *t*-test or ANOVA is used to analyse the differences among the means of these groups.

$$H_O: \mu_1 = \mu_2 = \cdots = \mu_n$$
$$H_A: \mu_1 \neq \mu_2 \neq \cdots \neq \mu_n$$

The method assumes that each sample is drawn from a normal population; each sample has the same variance.

$$F = \frac{\text{between groups estimated variance}}{\text{within groups estimated variance}}$$

Sample variance:

$$S^2 = \frac{\sum (x - \bar{x})^2}{n - 1}$$

Variance among samples means:

$$S_{\bar{x}}^2 = \frac{\sum (\bar{x} - \bar{\bar{x}})^2}{k - 1}$$

where: $\bar{\bar{x}}$ = the ground mean (i.e. the arithmetic mean of all the values of all the samples)
k = the number of samples.

As the standard error of the mean, $\sigma_{\bar{x}}$ is $\sigma / \sqrt{n}$, Levin and Rubin (1990, p. 439) show that the first estimate of the population variance, the between-groups variance, is:

$$\hat{\sigma}^2 = \frac{\sum n_j (\bar{x}_j - \bar{\bar{x}})^2}{k - 1}$$

n_j = number of items in sample j.

The within-group variance:

$$\text{sample variance, } S^2 = \frac{\sum (x - \bar{x})^2}{n - 1}$$

The second estimate of the population variance, the within-group variance, is:

$$\hat{\sigma}^2 = \sum \left(\frac{n_j - 1}{n_T - k} \right) S_j^2$$

where:

$$n_T = \sum n_j$$

As $F \to 1$, the likelihood that H_O is valid increases; as the value of F increases, the likelihood of H_O being valid decreases.

Degrees of freedom in the numerator: $(k - 1)$
Degrees of freedom in the denominator: $(n_T - k)$

Using tables of the F-distributions and the appropriate degrees of freedom; if $F_{calc} < F_{tab}$, the null hypothesis should not be rejected.

To be valid, the F-test can be applied to large samples only, $n \geq 100$ (Yeomans 1968, p. 101).

When research designs involve more than two conditions (therefore involving more than two means), it is better to analyse the data using ANOVA rather than many *t*-tests in order to reduce the chances of Type I error. ANOVA analyses differences between all condition means simultaneously. Rather than testing the difference between each pair of means as a *t*-test does, ANOVA determines whether any of a set of means differs from another using a single statistical test regardless of how many group means are involved in the test.

Example
Independent groups

Condition	Groups	Mean
A	1	μ_1
B	2	μ_2
C	3	μ_3
D	4	μ_4
E	5	μ_5

Instead of conducting 10 *t*-tests (i.e. groups 1 vs. 2, 1 vs. 3, 1 vs. 4 and so on) among all pairs of five means (each *t*-test with 5% chance of Type I error), ANOVA performs a single, simultaneous test on all condition means with 0.05 chance of Type I error.

In factorial designs, the between-groups variance can be broken down further into other components to test for the presence of different main effects and interactions, i.e. when the research design involves more than one independent variable, ANOVA can test whether the between-group variance is related to each of the independent variables, as well as whether the between-group variance is produced by interactions among the variables.

In a 2×2 factorial design involving two variables, α and β, the total variance would come from (1) the error variance, (2) the main effect of α, (3) the main effect of β, (4) the $\alpha \times \beta$ interaction (combined effect of the two variables). F-values are calculated for the main effect of α, β and $\alpha \times \beta$ interaction. Each of these F-values is then compared to the critical value of F. When an F-test is statistically significant, i.e. when the calculated value of F exceeds the critical value, at least one of the group means differs from one of the others. If the calculated F-value is less than the critical value, the null hypothesis is not rejected and so, conclude that the condition means do not differ.

Example
Factorial design

	Variables	α_1	α_2
Variables β_1 β_2			

The two conditions of variable β are absence and presence of anti-frost additive and the two conditions of variable α are ambient temperature levels of $+4$ and $-4°C$. Concrete cube strength is then tested while holding other variables constant, e.g., water at $4°C$.

When the independent variable has three levels or more, a significant F-test does not always tell us which means differ because ANOVA tests all condition means simultaneously, e.g. perhaps all of the means differ from each other, or only one mean differs from the rest. Hence, if ANOVA reveals a significant effect for an independent variable that has only two levels, no further tests are necessary because the F-test tells us that the two means differ significantly. However, if the independent variable has more than two levels, further tests are required, i.e. when a difference exists between at least two of the three condition means, but it does not indicate which means differ from the other. To identify which means differ significantly, researchers use follow-up tests, e.g. least significant difference (LSD) test, Tukey's test etc. Although differing in specifics, each of these tests is used after a significant F-test to determine precisely which condition means differ.

Regression and correlation

Usually, regression and correlation are considered together in expressing a relationship between two variables: one or more known values, realisations, of the independent variable; and the other unknown, the dependent variable. To keep research clear, it is advisable, at least in the beginning, to consider variables in pairs – one independent and the other dependent.

Regression and correlation statistics establish only any *relationship* between the realised values of the variables which occur; they *do not establish causality*, that is the province of theory, evidence and logical reasoning, in the light of the statistics. Conventionally, the independent variable is plotted on the x-axis and the hypothesised dependent variable on the y-axis. Simple or linear regression considers straight line hypothesised relationships only.

The standard form equation for a straight line is:

$$y = a + bx$$

where a is the intercept of the line on the y-axis and b is the slope of the line.

So, given at least two data points on a scatter plot (a graph of the associated values of x and y), a regression line can be drawn.

Example

Consider the following scatter plot:

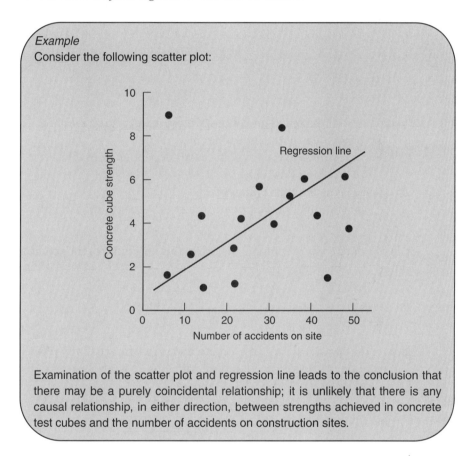

Examination of the scatter plot and regression line leads to the conclusion that there may be a purely coincidental relationship; it is unlikely that there is any causal relationship, in either direction, between strengths achieved in concrete test cubes and the number of accidents on construction sites.

Note, however, that there is a positive relationship when the regression line has a positive slope, upwards from left to right.

The regression line, the line of 'best fit' through the data points, uses the criterion of least squares. Squaring the vertical distance of each data point from the regression line both magnifies errors and removes the possible can-celling effects of positive and negative distances. A regression line is used for estimation – there will be errors between the line and the actual, realised data points. As the line is used to estimate points on the y-axis, it is usual for the equation for a straight line of estimation to be:

$$\hat{y} = a + bx$$

where: $\hat{y}$ (y-hat) are values on the y-axis estimated by the equation.

Least-squares error, to determine the line of best fit, minimises $\sum(y - \hat{y})^2$.

Given a set of data points which relate the independent variable x and the (hypothesised) dependent variable y, Levin and Rubin (1990, p. 491) note the equations to find the line of best fit to be:

$$b = \frac{\sum xy - n\bar{x}\bar{y}}{\sum x^2 - n\bar{x}^2}$$

$$a = \bar{x} - b\bar{y}$$

where: b = slope of the line of best fit (estimate/regression) line
x = values of the independent variable
y = values of the (hypothesised) dependent variable
$\bar{x}$ = mean of the values of x
$\bar{y}$ = mean of the values of y
n = number of data points (pairs of values of the variables x, y)
a = y-intercept.

The standard error of estimate measures the variability of the actual (realised) values from the regression line.

$$S_e = \sqrt{\frac{\sum(y - \hat{y})^2}{n - 2}}$$

Hence, the standard error of estimate measures the reliability of the estimating equation; analogous to standard deviation, it is a measure of dispersion. Levin and Rubin (1990, pp. 498–500) note:

$$S_e = \sqrt{\frac{\sum y^2 - a\sum y - b\sum xy}{n - 2}}$$

Assuming a normal distribution applies, the standard error of estimate exhibits the same properties as standard deviation, and can be used in the same way, for determining variability of predictions and confidence in them.

Regression assumes that the scatter of data points around the line of best fit is 'random', otherwise called *homoscedastic*. If there is a pattern to the scatter of the data points about the line, which shows the scatter to be different at different points, *heteroscedasticity* is present and so, regression is questionable. Strictly, homoscedasticity is where the error has constant variance;

heteroscedasticity is where the variance of the error changes along the length of the regression line (Pindyck & Rubinfeld 1981, p. 49).

Not all scatter plots suggest that straight lines are the best fit, so curve fitting may be appropriate. Fortunately, as for a straight line, there are standard forms of equation for various types of curves – usually determined by the approximate shapes of the curves, such as slope changes and any turning points. Use of computer packages is helpful for both linear and non-linear regression. The shape of the line and its nature should be detectable, along with any close alternatives, from observation of the scatter diagram and, in particular, from the nature of the relationship suggested by any underpinning theory. Theory is important!

Example
In the long term, the relationship between a person's age and her/his cash income is likely to be of the form shown below, whilst that person is of working age and in continuous employment.

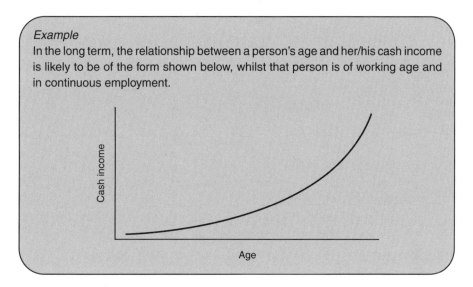

The line is an increasing exponential curve as income increases by an approximately constant percentage each year, so that the cash sum received per annum increases progressively. Economic theory has been used to determine the nature of the relationship. Collection of data and application of statistics would allow analytic checking and quantification of the relationship. Fortunately, most statistics packages, via simple commands, manipulations of data and setting of criteria, allow the program to determine the line of best fit.

The *coefficient of correlation, r,* identifies the degree and nature of the relationship between the two variables, from a perfect positive relationship (+1) to a perfect negative relationship (−1), i.e.

$$-1 \leq r \leq +1$$

$r = +1$ means that an increase in variable x is matched by an equiproportional increase in y. If $r = 0$ there is no relationship; changes in the variables are quite independent of each other; they are random. However, it is common to wish to know how much of the change in the values of a dependent variable are caused, given the logic of the relationship, by a change in the values of the independent variable. The statistic required is the *coefficient of determination*, r^2.

If $r^2 = 0.81$ (i.e. $r = 0.9$), 81% of the changes in y are caused (explained) by the changes in x.

The coefficient of determination can be calculated using the following method:

A sample of data has variation about its own mean which, in terms of least-squares error, is:

$$\sum (y - \bar{y})^2$$

Similarly, the variation of the data about the regression line is:

$$\sum (y - \hat{y})^2$$

The sample coefficient of determination is:

$$r^2 = 1 - \frac{\sum (y - \hat{y})^2}{\sum (y - \bar{y})^2}$$

The coefficient of determination measures the strength of a *linear* relationship between two variables.

Levin and Rubin (1990, p. 510) note a 'short-cut method' for calculation of r^2 and most statistics packages can calculate r^2 directly:

$$r^2 = \frac{a \sum y + b \sum xy - n\bar{y}^2}{\sum y^2 - n\bar{y}^2}$$

If the true regression line for the population is given by $y = a + bx$ and the line estimated from the sample is $\hat{y} = a + bx$, the standard error of the regression coefficient can be used to test null hypotheses – e.g. that the slope of the regression line is unchanged from what has been found in the past (the 'proportionate' relationship between y and x). This is analogous to the use of standard deviation.

As b denotes the slope of the population's regression line, S_b denotes the standard error of the regression coefficient of b.

$$S_b = \frac{S_e}{\sqrt{\sum x^2 - n\bar{x}^2}}$$

$$\left(S_e \text{ is the standard error of estimate; } = \sqrt{\frac{\sum y^2 - a \sum y - b \sum xy}{n - 2}} \right)$$

Using the t-distribution with $n - 2$ degrees of freedom, the limits for the acceptance region in this instance, showing that b is unchanged, are:

Upper: $b + t(s_b)$
Lower: $b - t(s_b)$

Confidence intervals can be calculated in a similar way.

For some types of data, such as opinion surveys, which have employed Likert scales or something similar, the data are not suitable for analysis by regression and correlation due to the nature of the scales employed.

Example

Data on satisfaction with projects procured by various approaches have been collected from clients and contractors using a 5-point Likert scale (1-totally satisfied, 5-totally dissatisfied). The data are summarised below. All cell figures are numbers of respondents scoring per cell.

Procurement method	Client Satisfaction					Contractor Satisfaction				
	1	2	3	4	5	1	2	3	4	5
Traditional	2	4	7	6	3	1	5	10	0	4
Construction management	3	5	1	8	2	5	5	5	3	2
Management contracting	4	6	2	8	0	6	6	5	2	1
Project management	2	4	6	2	6	3	7	3	4	3
Design & build	5	2	6	3	4	6	6	5	1	2
Design & management	2	3	7	4	4	5	5	3	5	2
BOOT*	1	6	7	3	3	6	6	4	2	2

*The procurement method: Build, Own, Operate, Transfer.

Means and standard deviations are not appropriate. Rank correlations should be used as only the rankings can be compared. Rankings of satisfaction are obtained from examining the scorings in the totally satisfied column (column 1) – only in the event of tied scores are the next (and then subsequent) columns considered to determine the ranks.

	Clients	Contractors	D (Difference)	D^2
Traditional	4	7	3	9
Construction management	3	4	1	1
Management contracting	2	1	1	1
Project management	5	6	1	1
Design & build	1	2	1	1
Design & management	6	5	1	1
BOOT	7	3	4	16
				30

The approach is adopted because respondents interpret the scorings differently in terms of both levels of the score categories (1–5 and associated word descriptors, if any) and of the intervals between the score categories. The latter differences may be accentuated if the data are collected against word descriptors only.

The coefficient of correlation between the ranks is a measure of the association between two variables which is determined from the ranks of observations of the variables. It is calculated using Spearman's coefficient of rank correlation, ρ:

$$\rho = 1 - \frac{6 \sum D^2}{n(n^2 - 1)} = 1 - \frac{6 \times 30}{7(7^2 - 1)}$$

$$= 1 - \frac{180}{7 \times 48} = 1 - 0.54 = \underline{0.46}$$

As Likert scales yield ordinal data, strictly, regression and correlation cannot be used, as those analytic techniques require interval or ratio data. However, following Labovitz (1970), especially where ordinal variables permit a large number of categories to be specified, the variables can be treated as interval data, especially as techniques like regression and correlation are well known, powerful and quite easy to use and interpret. The view is controversial and so, before treating ordinal data as interval data, advice should be obtained from an expert statistician regarding the validity of such an adaptation for the particular data: if in doubt, be strict in the treatment of data.

Multiple regression

Regression analysis refers to relations of changes in levels of y to changes in levels of x. In multiple regression, the value of the predicted outcome variable y is viewed as depending on, the intercept on the y-axis, and the values of the predictor variables x_1, x_2, x_3, x_k etc. multiplied by a coefficient β chosen in practice so as to minimise the sum of the squared discrepancies between the predicted and obtained values of y. A term c is added to describe the discrepancy between a particular value of y and the predicted value for that y. Thus, for two predictor variables, x_1 and x_2, the equation is:

$$y = \alpha + \beta_1 x_1 + \beta_2 x_2 + c$$

As the number of predictor variables increases, all the β's and α's change so that the magnitude, sign and statistical significance of each regression coefficient depends entirely on exactly which other predictor variables are in the regression equation.

> *Example*
> A multiple regression model is developed to predict the average hourly mainte-
> nance cost of tracked hydraulic excavators operating in the UK open cast mining
> industry. The performance of this model is then compared to an artificial neural
> network (ANN) model. (See Edwards *et al.* (2000) for details.)
>
> A multiple regression model is used to relate bidder competitiveness (the
> dependent variable) to the independent variables of bidder, contract type and
> contract size. (See Drew & Skitmore (1997) for details.)

A special problem of multiple regression is that of collinearity, or high
correlations among the predictor variables, e.g. age and experience as fellow
predictor variables. Collinearity makes it hard to interpret the substantive
meaning of regression coefficients. Moses (1986) points out that one consequence of collinearity is that we may have a large R^2 and yet find none of the
regressors to be significant.

Canonical correlation is the correlation between two or more predictor
variables and two or more dependent variables, The canonical correlation
coefficient is a Pearson product–moment correlation between a composite-
independent and a composite-dependent variable. Rosenthal and Rosnow
(1991) do not recommend canonical correlation for hypothesis testing or
confirmatory analyses, but find that the procedure is useful from time to
time as an hypothesis-generating procedure, i.e. in the spirit of exploratory
data analysis. 'For the situation for which canonical correlations apply, we
have found it more useful to generate several reasonably uncorrelated independent supervariables (with the help of principal components or cluster
analysis) and several reasonably uncorrelated dependent supervariables (with
the help of principal components or cluster analysis)' (Rosenthal & Rosnow
1991, p. 560).

Time series

Most of the data which are used by researchers are time series. These are
measurements of a continuous variable, such as temperature of the air at a
particular location, made at constant intervals of time over a period. As the
measurements are instantaneous representations of a variable which changes
continuously, joining the points on a plot produces a graph of the time series.
   Time series have four component parts:

- Secular trend (T)
- Cyclical fluctuation (C)

- Seasonal variation (S)
- Residual component irregular/random variation (R).

Quite simple techniques can be employed to break down the time series into its deterministic components; however, certain aspects must be considered first. The *nature* of the model – the way in which the components aggregate to produce the realisations – is either *additive* or *multiplicative* (i.e. $A = T + C + S + R$; or $A = T \times C \times S \times R$). The relevance of the data must be evaluated – annual data cannot reveal a seasonal component, and short runs of data cannot reveal cycles. In economics, cycles are short, medium and long; long cycles may be of around 50 years.

If the data cannot reveal certain components, the *hidden* component is considered to be a joint part of the residual.

Having collected realisations of, if possible, raw data, the first step is to produce a scatter plot, then the line of best fit is used to represent the secular trend. Often, line-fitting methods, using the regression techniques described earlier, are used to determine the line of best fit. The usual criterion for determining the line of best fit is minimum least-squares error. For long duration runs of data, it is important not to fit a secular trend line of too complex a shape such that it will absorb some seasonal/cyclical components. It is best to select a line of quite simple shape and well-known mathematical formula – straight line, logarithmic exponential, Gompertz, logistic, polynomial. A good criterion is to use the simplest, appropriate form of line; this approach is called 'parsimony'.

The standard mathematical equations for the lines just noted are:

straight:	$y = a + bx$
logarithmic:	$y = ab^x$
exponential:	$y = ae^x$ (a 'special case' of logarithmic)
Gompertz:	$y = ka^b$
logistic:	$1/y = k + ab^x$
polynomial:	$y = a + bx + cx^2$; $a + bx + cx^2 + dx^3$ etc.

Familiarity with the shapes of the lines produced by the standard form equations will be a notable help in trend-fitting. It is usually better to avoid high powers of x.

Alternative methods to statistical line fitting to represent the secular trend include *semi-averages* and *moving averages*. For semi-averages, the data set is divided into halves, the average of each half-set is calculated and the semi-average trend line is the straight line through the two half-set average points, i.e. the semi-average points. For moving averages (MA), an appropriate number of data points must be selected for averaging. The selection depends on the nature of the data, the periods of sampling, and the amount of 'smoothing' desired, since the more the data points are averaged, the greater the smoothing. Using an odd number of data points, the moving average is 'centred' on a sampling time automatically. However, for a moving average with an even number of data points, weightings must be used to 'centre' the averages on sampling times.

Example
Assume quarterly data; i.e. sampling is at 3-monthly intervals. In this instance, averaging is over three quarters.

Quarter	Data	Total for 3 Quarters	3QMA
1	10		
2	11		11.3
3	13	34	12.7
4	14	38	13.7
5	14	41	14.7
6	16	44	15.0
7	15	45	16.3
8	18	49	

To calculate the 4QMA (common for quarterly data) from the last table of data, the formula to centre the moving averages on sampling times is:

$$\frac{A_{t-2} + 2A_{t-1} + 2A_t + 2A_{t+1} + A_{t+2}}{8}$$

A = actual data point
A_t = actual data value at time t.

The moving average is centred on A_t with the preceding and succeeding quarterly data at distance two quarters given a single weighting each, other data points being given a double weighting each, to preserve a balanced centring.

Example

Quarter	Data	Weighted 4Q Total	4QMA
1	10		
2	11		
3	13		12.5
4	14		13.62
5	14	100	14.5
6	16	109	15.25
7	15	116	
8	18	122	

In order to fit a straight line ($y = a + bx$) to a set of sample data:

$$a = \bar{y} - b\bar{x} \quad \text{and} \quad b = \frac{\sum xy - n\bar{x}\bar{y}}{\sum x^2 - n\bar{x}^2}$$

For the sample data, 'rounded' about the mid-point of x, i.e. $\bar{x} = 0$ (see table below), this reduces to $a = \bar{y}$ and $b = \sum xy / \sum x^2$.

Year/ Qtr		(Data; A) Output – Const 1975 Prices [y]	Round Q1 1977	Round Q1 1978/ Q4 1979 [x]	x^2	xy	Linear Trend (T) $y_t = 2745 + 7.28x$
1977	1	2511	0	−15	225	−37 665	2636
	2	2612	2	−13	169	−33 956	2650
	3	2698	4	−11	121	−29 678	2665
	4	2654	6	−9	81	−23 886	2679
1978	1	2562	8	−7	49	−17 934	2694
	2	2854	10	−5	25	−14 270	2709
	3	2910	12	−3	9	−8730	2723
	4	2804	14	−1	1	−2804	2738
1979	1	2523	16	1	1	2523	2752
	2	2848	18	3	9	8544	2767
	3	2951	20	5	25	14 755	2781
	4	2908	22	7	49	20 356	2796
1980	1	2764	24	9	81	24 876	2811
	2	2809	26	11	121	30 899	2825
	3	2858	28	13	169	37 154	2840
	4	2648	30	15	225	39 720	2854
		43 914		0		9904	

The best prediction of the series shown in the last data column is the trend plus seasonal components. The sample data were insufficient to yield a cyclical component and the residuals are random. Given appropriate analysis of the other components, over time, the mean of the residuals is zero.

Note: In the example, the additive model where $A = T + C + S + R$, was used; in practice (especially in economics), the multiplicative model is preferable. No cyclical fluctuation can be calculated in this example as four years is not sufficient data hence, the cyclical fluctuation component is absorbed by each of the three other components. Hence, in the table on the next page, the residual component is calculated as $R = A - T - S$.

Year/ Qtr		Linear Trend (T)	Detrended Series (A − T)	Seasonal Component (S)	Seasonally Adjusted Series (A − S)	Residual Component (R)	Predicted Series
1977	1	2636	−125	−133	2644	8	
	2	2650	−38	43	2569	−81	
	3	2665	33	103	2595	−70	
	4	2679	−25	−13	2667	−12	
1978	1	2694	−132	−133	2695	1	
	2	2709	145	43	2816	102	
	3	2723	187	103	2807	84	
	4	2738	66	−13	2817	79	
1979	1	2752	−229	−133	2656	−96	
	2	2767	81	43	2805	38	
	3	2781	170	103	2848	67	
	4	2796	112	−13	2921	125	
1980	1	2811	−47	−133	2897	86	
	2	2825	−16	43	2766	−59	
	3	2840	18	103	2755	−85	
	4	2854	−206	−13	2661	−193	
1981	1	2869		−133			2736
	2	2883		43			2926
	3	2898		103			3001
	4	2912		−13			2899
1982	1	2927		−133			2794
	2	2842		43			2885

Note: $\sum R \approx 0$.

The detrended series is used to construct a table, as shown below, so that the seasonal components of the time series can be calculated.

Year	1Q	2Q	3Q	4Q	
1977	−125	−38	33	−25	
1978	−132	145	187	66	
1979	−229	81	170	112	
1980	−47	−16	18	−206	
Total	−533	172	408	−53	[−6]
Corrected: [−6/4 = −1.5]	−531	173	410	−52	
Average [÷4]	−133	43	103	−13	[To nearest whole number]

Index numbers

Index numbers are a means of measuring changes of a *composite* entity over time which, itself, is not quantifiable directly, but the components of which may be measured and aggregated (an index number is analogous to a construct which is made up of several variables). Common examples of index numbers are the retail prices index and the FT100 share index. The composite nature of an index number distinguishes it from a *price* or *quantity relative*. A quantity relative is the quantity of an item in the current year compared with the quantity in the base year (e.g. quantity bought in these years described as a number of units bought). A price relative is the price of an item in the current year compared with the price in the base year (usually price per unit). Index numbers use the concept of a 'basket of goods', which is a composite entity, to provide the weightings to be applied to price relatives.

Example *Price relative*

Year	Price/unit (£)	Index
1950 (base year)	20.00	100
1955	24.00	120
1960	29.00	145
1965	34.00	170
1970	40.00	200
1975	47.00	235
1980	58.00	290
1985	70.00	350

Index numbers are used extensively in published statistics. They show the change from a base point of index 100. Analysis of changes between points, other than with the base, requires an element of arithmetic. A further element of possible difficulty or confusion is that if two series of index numbers, say a price index – PI – and a cost index – CI, have the same base point and both are allocated 100 at that point, e.g. 1 January 1990, it does *not* mean that prices and costs were equal at 1 January 1990; subsequent proportional changes in the two index series can be considered.

Example

If at 1 April 1991, PI = 120 and CI = 115, prices have increased, on average, 20% over the 15-month period whilst costs have increased by 15%. Thus, over the 15-month period, the activity became more profitable – by 5% of the base price.

Simple average index

The simple average index takes no account of the different units of sale, patterns of consumption and changes in those patterns.

Example

Item	Price/unit 1960 (pence)	Price/unit 1985 (pence)
Bricks	20	40
Electric cable	35	60
Sand	25	50
PVC pipe	30	40
Cement	60	100
Aggregate	20	30

$$I = \frac{40 + 60 + 50 + 40 + 100 + 30}{20 + 35 + 25 + 30 + 60 + 20} \times \frac{100}{1}$$

$$= \frac{32\,000}{190}$$

$$= \underline{\underline{168}}$$

Production of index numbers comprises relating ratios which express the change in the price or quantity to a particular parameter – usually time. Price relatives are expressed for a predetermined quantity. Normally, determination of price relatives involves sampling; the average price relative is used for the index. The variability of the price relatives for the goods in the basket about the average is important in establishing and maintaining the validity of the index.

Components of an index:

Quantities in year

$$\text{Prices in year} \begin{bmatrix} 0 & \sum p_0 q_0 & \sum p_0 q_1 & \sum p_0 q_2 \\ 1 & \sum p_1 q_0 & \sum p_1 q_1 & \sum p_1 q_2 \\ 2 & \sum p_2 q_0 & \sum p_2 q_1 & \sum p_2 q_2 \end{bmatrix}$$

$p_0, p_1, p_2 = $ prices in years 0, 1 and 2
$q_0, q_1, q_2 = $ quantities in years 0, 1 and 2.

The main diagonal, or trace, of the matrix (top left to bottom right) gives measures of value of the constituents of the index.

$$V_{01} = \frac{\sum p_1 q_1}{\sum p_0 q_0} \qquad V_{02} = \frac{\sum p_2 q_2}{\sum p_0 q_0}$$

V_{01} = change of value from year 0 to year 1.

The relationship for the change in value from year 1 to year 2 is:

$$V_{12} = \frac{\sum p_2 q_2}{\sum p_1 q_1}$$

If the index comprises a variety of items, i, where $i = 1, 2, 3, \ldots, n$, then

$$\sum_{i=1}^{n} p_{i0} q_{i0} = \sum p_0 q_0$$

A 2-case matrix:

$$\text{Prices} \begin{array}{cc} & \begin{array}{cc} \text{Quantities} \\ 0 \qquad\qquad 1 \end{array} \\ \begin{array}{c} 0 \\ 1 \end{array} & \begin{bmatrix} \sum p_0 q_0 & \sum p_0 q_1 \\ \sum p_1 q_0 & \sum p_1 q_1 \end{bmatrix} \end{array}$$

Base weighted and *current weighted* indices are the most common standard forms of index numbers.

The *base weighted* (Laspèyres) indices are:

$$\text{Price:} \quad \frac{\sum p_1 q_0}{\sum p_0 q_0} \quad \text{i.e. } P_{01}(q_0)$$

$$\text{Quantity:} \quad \frac{\sum p_0 q_1}{\sum p_0 q_0} \quad \text{i.e. } Q_{01}(p_0)$$

Laspèyres indices are base weighted (usually at year 0). As the weighting applies to the run of index numbers thereafter, the weightings are *fixed* (as at the base).

The *current weighted* (Paasche) indexes of price and quantity are:

$$\text{Price:} \quad \frac{\sum p_1 q_1}{\sum p_0 q_1} \quad \text{i.e. } P_{01}(q_1)$$

$$\text{Quantity:} \quad \frac{\sum p_1 q_1}{\sum p_1 q_0} \quad \text{i.e. } Q_{01}(p_1)$$

The first row of the matrix denotes expenditure at constant prices at the base year; columns represent expenditures at constant quantities (of the particular years). Quantity indexes represent expenditures in 'real terms'.

Laspèyres indexes are both *fixed weighted* and *base weighted*. It is usual for the reference base (year) to be the year used for the weight's base. Most published indexes are rebased periodically; this helps to maintain their validity.

A Laspèyres price index, $P_{0t}(q_0)$ can be rebased whilst retaining its fixed weighting by switching rows:

$$P_{1t}(q_0) = \frac{P_{0t}(q_0)}{P_{01}(q_0)} = \frac{\sum p_t q_0}{\sum p_1 q_0}$$

It can retain its base weighting but obtain a different reference base by switching columns:

$$P_{1t}(q_t) = \frac{\sum p_t q_1}{\sum p_1 q_1}$$

The construction of some simple index numbers occurs as follows:

Example

Weekly Wage Rates (£)

Employee Type	1950	1955	1960
Unskilled	8.75	10.75	14.50
Semi-skilled	9.75	12.50	16.00
Skilled	12.00	14.00	18.50
Clerical	10.00	12.25	13.00
Total	**40.50**	**49.50**	**62.00**

Weekly Wage Rate Relatives, 1950 = 100

Employee Type	1950	1955	1960
Unskilled	100	122.9	165.7
Semi-skilled	100	128.2	164.1
Skilled	100	116.7	154.2
Clerical	100	122.5	130.0
Total	**400**	**490.3**	**614.0**

Index of weekly wage rates:

$$1950: \quad \frac{400}{4} = 100.0$$

$$1955: \quad \frac{490.3}{4} = 122.6$$

$$1960: \quad \frac{614.0}{4} = 153.5$$

By using aggregated wage rates:

Aggregates: 1950 = 40.50; 1955 = 49.50; 1960 = 62.00

$$1950: \quad \left(\frac{40.50}{4} \div \frac{40.50}{4}\right) \times \frac{100}{1} = \frac{40.50}{40.50} \times \frac{100}{1} = 100$$

$$1955: \quad \frac{49.50}{40.50} \times \frac{100}{1} = 122.2$$

$$1960: \quad \frac{62.00}{40.50} \times \frac{100}{1} = 153.1$$

	Assume Constant 1950–1960	
	(Quantity Weights)	**(Value Weights 1950)**
	No. of Employees	**Average Wages Bill (£)**
Unskilled	9	102.00
Semi-skilled	23	293.25
Skilled	17	252.17
Clerical	1	11.75
Total	**50**	**659.17**

Relatives	Wage Rates Relative to:						
	Average Wages Bill	1950	1955	1960			
	(1)	(2)	(3)	(4)	(1) × (2)	(1) × (3)	(1) × (4)
Unskilled	102.00	100	122.9	165.7	10 200	12 535.800	16 901.40
Semi-skilled	293.25	100	128.2	164.2	29 325	37 594.650	48 151.65
Skilled	252.17	100	116.7	154.2	25 217	29 428.239	38 884.63
Clerical	11.75	100	122.5	130.0	1175	1439.375	1527.50
Total	**659.17**				**65 917**	**80 998.064**	**105 465.18**

$$1950: \quad \frac{65\,917}{65\,917} = 100.00$$

$$1995: \quad \frac{80\,998.064}{65\,917} = 122.9$$

$$1960: \quad \frac{105\,465.18}{65\,917} = 160.0$$

Example

Consider the following:

Year	Index (1955 = 100)
1960	119.0
1961	121.3
1962	125.0
1963	126.4
1964	128.0
1965	132.0
1965	100.0 (1965 = 100)
1966	102.1
1967	104.0

To convert the second run to 1955 = 100 base:

$$1965: \quad 132.0 \times \frac{100.0}{100.0} = 132.0$$

$$1966: \quad 132.0 \times \frac{102.1}{100.0} = 134.8$$

$$1967: \quad 132.0 \times \frac{104.0}{100.0} = 137.3$$

Using numbers of employees in each category and their wage rates yield index numbers of:

1950:	100.0
1955:	122.7
1960:	159.8

Hence, the method of producing the indexes does (if only marginally) affect the result.

Frequently, long series of index numbers have the base changed periodically; the manipulation required to convert to a common base is quite straightforward.

Chained index

Often, short runs of index numbers are produced which must be 'spliced', i.e. joined together coherently, to produce long runs. This is similar to the change of base calculation.

An alternative is to produce a chain index. Although, ideally, a chain index would be updated continuously for changes in quantities as well as prices, in practice, for indexes such as the Retail Prices Index (RPI), *chaining* i.e. adjusting quantities etc., occurs annually.

Example

Year	GDP at 1958 Constant Prices (£m)		Index	Price Index 1958 = 100
1956	21 070	$\dfrac{21\,070}{21\,478} \times \dfrac{100}{1} =$		98.1

Year	GDP at 1958 Constant Prices (£m)		Index	Price Index 1958 = 100
1957	21 474	$\dfrac{21\,474}{21\,070} \times \dfrac{100}{1} =$	101.9	100.0
1958	21 478	$\dfrac{21\,478}{21\,474} \times \dfrac{100}{1} =$	100.0	100.0
1959	22 365		104.1	104.1
1960	23 484		105.0	109.3
1961	24 268		103.3	113.0
1962	24 442		100.7	113.8
1963	25 537		104.5	118.9
1964	26 912		105.4	125.3

Index numbers are a convenient way of representing a time series to demonstrate relative, proportional changes from the base. Beware of changes occasioned by sampling variations, splicing, chaining etc. Remember that index numbers do *not* give absolute figures.

As it is common for index numbers to be used to depict time-series data, it is appropriate to reinforce the wisdom by drawing a diagram of the data in as 'raw' a state as possible. An important aspect of time series is that the realisations are subject to a constant set of probability laws – this is a vital consideration for predictability; an essential of management. Plotting the data makes it easy to identify the likelihood of the probability requirements being met and, often more obviously, any incidences of 'discontinuities' or *shocks* – external (*exogenous*) influences which cause a disruption to the 'smooth' flow of the data stream.

Shocks are important but, due to their nature, cannot be predicted. In his major work, *Building Cycles and Britain's Growth*, Parry-Lewis (1965) considered the most important influences on levels of construction activity to be:

- population
- interest rates, and
- shocks

Other analytical techniques

Cluster analysis

Appropriate algorithms (mathematical rules or procedures) are used in cluster analysis to split the data into clusters/groups. There are two basic types of clustering technique: *hierarchical* and *partitioning* (as performed by the Statistical Package for Social Sciences (SPSS) – see SPSS 1997). The data are sorted based on optimising some predefined criteria (Dillon & Goldstein 1984). Whilst the hierarchical method performs successive division of the data, which produces irrevocable allocation of clusters, the partitioning method allows data to switch cluster membership. Hierarchical algorithms make one pass through a data set and, therefore, poor cluster assignment cannot be modified (Ketchen & Shook 1996). In the partitioning method, by making multiple passes through the data, the final solution optimises within-cluster homogeneity and between-cluster heterogeneity – provided that the number of clusters is specified *a priori* (Ketchen & Shook 1996). However, according to Aldenderfer and Blashfield (1984), Ward's method (1963) is the most widely used cluster method in social sciences and has been shown to outperform other cluster procedures.

Common methodological issues in the use of cluster analysis involve (1) the selection of the number of clusters and (2) testing for differences among clusters. Major jumps in fusion coefficients at each agglomerative stage can be examined to show the number of clusters; i.e. the jump indicates the suggested 'cut-off' (see Ulrich & McKelvey (1990) e.g.). To test for the differences among clusters, the within-group distance can be compared with the across-group distance.

Example
Sabherwal and Robey (1995) seek to classify the sequences of events that affect the information system (IS) implementation processes in 53 organisations. Cluster analysis is used to develop an empirical taxonomy of IS implementation procedures based on intersequence distances and each sequence of events represents one data point.

continued

> *Example   Continued*
>   To test for differences among clusters: (1) compute the mean distance of each sequence from the other sequences within its cluster; (2) for each cluster, consider the sequence with the smallest mean distance from the other members of the cluster to make an approximation to the cluster centroid; (3) perform *t*-test or *F*-test to compare the mean distance of each sequence from the other sequences within its cluster with its mean distance from the sequences approximating centroids of the other clusters.
>
>   Comparison of the mean distance from the other sequences within the same cluster with the mean distance from the sequences approximating cluster centroids of the other clusters produced a *t*-statistic of 2.42, significant at the 0.01 level. This result indicates that the within-group distances are less than the across-group distances, and that the *clusters are distinct.*

Factor analysis

Factor analysis is a multivariate method which analyses relationships among difficult to interpret correlated variables in terms of a few conceptually meaningful, relatively independent factors, each of which represents some combination of original variables (Rummel 1970; Kleinbaum *et al.* 1988; Comrey & Lee 1992), i.e. variables are grouped into a relatively small number of factors (factor extraction) that can be used to represent relationships among sets of many interrelated variables (see Norusis 1992).

Thus, a factor is a type of *latent construct* in that a construct is an amalgamation of variables and is latent because it cannot be observed (and measured) directly but only through the constituent variables. Usually, the goodness of fit of the variables which are believed to combine to constitute the construct is assessed by means of calculating *Chronbach's alpha* (α), which is a coefficient of reliability (consistency). Thus, Chronbach's alpha uses metrics of the number of variables which are believed to be the constituents of the construct and the correlations between them:

$$\alpha = \frac{N\bar{r}}{1 + (N-1)\bar{r}}$$

where: N is the number of variables

   $\bar{r}$ is the average inter-variable correlation of all the variables.

Generally, the critical level for reliability when using Chronbach's alpha is 0.7; any coefficient below that indicates that the variables are not sufficiently inter-correlated to combine to yield a single latent construct. Apart from being used in factor analysis, Chronbach's alpha is used widely as a reliability indicator – such as in developing scales of measurement in the Social Sciences.

Usually, factor extraction is done by means of principal components analysis which transforms the original set of variables into a smaller set of linear combinations that account for most of the variation of the original set. The principal components are extracted so that the first principal component accounts for the largest amount of the total variation in the data. The mth principal component $PC_{(m)}$ is that weighted linear combination of the observed variables X,

$$PC_{(m)} = w_{(m)1}X_1 + w_{(m)2}X_2 + \cdots + w_{(m)p}X_p$$

which has the largest variance of all linear combinations that are uncorrelated with all of the previously extracted principal components. Various tests are required for the appropriateness of the factor extraction, including the Kaiser–Meyer–Olkin (KMO) measure of sampling accuracy and the Barlett test of sphericity which tests the hypothesis that the correlation matrix is an identity matrix. Normally, a data set of at least 100 usable returns is necessary for factor analysis.

Since the distinctive characteristic of principal components analysis is its data-reduction capacity, it must determine the number of factors to be retained. Kaiser (1958) suggests that one criterion for determining the number of retained factors is to exclude factors with variances less than one. The rationale for this is that any factor should account for more variance than any single variable in the standardised test score space. Another approach is proposed by Cattell (1966), the 'scree test', where the eigenvalues of each component are plotted against their associated component. The scree plot helps to identify the number of factors to be retained by looking for a relatively large interval between eigenvalues. The rationale for the scree test is that since the principal component solution extracts factors in successive order of magnitude, the substantive factors appear before the numerous trivial factors which have small eigenvalues that account for a small proportion of the total variance. However, Dillon and Goldstein (1984) mention two complications of the scree test. First, there might be no obvious break, in which case the scree test is inconclusive. Second, in the case of having several breaks, it would be difficult to decide which break reflects the more appropriate number of factors.

Since the purpose of factor analysis is to group variables into factors (or principal components) determined by factor loadings, meaningful interpretation of the factors generated is important. Factor loadings (or coefficients) give the correlations between variables and factors. Whilst factor loading of 0.30 is often used as a cut-off for significance (so that variables with factor loadings of less than 0.30 are not included in the factor), Nunnally (1978, p. 434) suggests that it is doubtful that loadings smaller than 0.40 should be taken seriously: 'a . . . way to fool yourself with factor analysis is . . . to overinterpret the meaning of small factor loading . . .'. One may prefer the factor structure where the groups of variables are conceptually consistent and interpretable so that a factor label can be meaningfully assigned.

For ease of interpretation of the factor extraction, the principal components matrix is often rotated. There are several rotation methods available in SPSS and the more common ones are *varimax* and *oblimin*. Dillon and Goldstein (1984) assert that the varimax method is most popularly used to rotate principal components solutions. In simple terms, the procedure seeks to rotate factors so that the variation of the squared factor loadings for a given factor is made large to allow ease of interpretation based on the significance of the loadings.

Example

A comparative study is conducted on 84 UK contractors to determine the factors which influence contractors' cost estimating by means of factor analysis. KMO is 0.748, which is acceptable, and Barlett's test of sphericity is 977.239 with associated significance level at $p = 0.000$, suggesting that the population correlation matrix is not an identity matrix.

The variables are grouped into seven factors: project complexity, technological requirements, project information, project team requirement, contract requirement, project duration and market requirement.

For further details, see Akintoye (2000).

Path analysis

Path analysis is a generalisation of multiple regression that allows one to estimate the strength and sign of directional relationships for complicated causal schemes with multiple dependent variables (Li 1975). The result of path analysis is a model which explains the interaction of a large number of variables to illustrate the causality entertained in a network of relationships. The strengths of these relationships are measured by path coefficients which are standardised measures that can be compared to determine the relative predictive power of each independent variable with the effects of the other variables being partialled out.

Path analysis provides researchers with a multivariate approach (more than one dependent, endogenous variable which bring about simultaneous equations) to estimate, structurally, the direct, indirect and total causal effects among latent constructs – supposing that the theoretically sound model, covering *a priori* hypothesised causalities of the involved constructs, has been conceived (see Bollen 1989; Mueller 1996). The causal scheme is usually considered an *a priori* hypothesis of potential effects, and alternative hypotheses can be proposed and tested against each another. Conversely, the *a priori* causal scheme can be taken as a given and used to make predictions about patterns of evolution (Scheiner *et al.* 2000).

The particular value of path analysis is that it illustrates the working relationship of all variables in a network of relative predictive powers, thus allowing one to understand the relationships among variables in a systematic manner. Certain variables are singled out as 'causes' (exogenous variables) and other variables as 'effects' endogenous variables.

However, no statistical methodology is capable of establishing absolute cause and effect. 'Cause and effect relationships are derived from theory, and theory comes from outside of statistics' (Dillon & Goldstein 1984, p. 432). According to Nie *et al.* (1975, p. 397), if the main interest is in assessing the overall effect of one variable over another variable in the same sample, the standardised coefficients (path coefficients) are appropriate. However, if one is interested in finding causal laws or causal processes and/or in comparing parameters of one population with another, the unstandardised coefficients (structural coefficients) are preferred.

A path coefficient is the standardised slope of the regression of the dependent variable on the independent variable in the context of the other independent variables. If there is only a single independent variable, this standardised coefficient is a Pearson product–moment correlation; if there are additional independent variables, it is a standardised partial regression coefficient (Scheiner *et al.* 2000). In many situations, the path coefficients in path analysis turn out to be the same as the standardised beta coefficients in regression analysis. Thus, regression analysis is often used to build up causal models. 'Values for the path coefficients can be obtained from standard multiple regression computer programs' (Dillon & Goldstein 1984, p. 443). The SPSS program provides three main methods of regression analysis for building up the model by controlling the entry or removal of variables: forward selection, backward elimination and stepwise selection. Forward selection enters the variables into the model one by one with the strongest positive (or negative) simple correlation with the dependent variable and stops when an established criterion for the F no longer holds. Backward elimination begins with all candidate variables in the model, then, at each step, it removes the predictor that contributes least to the fit. Stepwise selection begins like forward stepping but, at each step, tests variables already in the model for removal. SPSS (1997) states that none of these procedures is guaranteed to provide the best subset in an absolute sense.

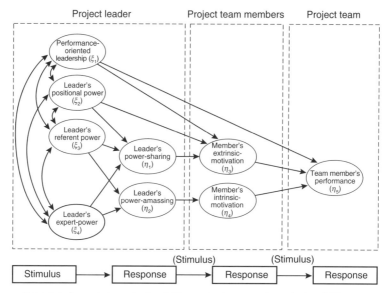

Note: ξ_1 to ξ_4 are independent variables and η_1 to η_5 are dependent variables. Unidirectional arrows indicate the theoretical causal relationships and the curved double arrows indicate the correlationship between the corresponding variables.

Fig. 7.5   A model of project leadership (adapted from Fang 2002).

A path diagram is a scheme of causal relationships (Fig. 7.5). The structured concepts can be drawn in a diagram to illustrate, simply, the structural (causal) relationships among them. A path model with only unidirectional linkage between each pair of constructs is referred to as the recursive path model.

Example

In Fig. 7.5, the ellipses are the abbreviations and symbols for latent constructs. The endogenous constructs refer to those that are dependent on other constructs within the model and are represented by the symbol η, whilst the exogenous constructs, represented by ξ, refer to those that are independent of effects, apart from influences from outside the model. The endogenous constructs are influenced by other endogenous constructs and/or the exogenous constructs. The latter are hypothesised as independent, since they are not causally affected by any constructs within the model. The error term associated with each endogenous construct (η_1) is represented by ζ_1. The error term can be considered as one special kind of explanatory (independent) factor which, together with the other explanatory (independent) constructs, takes account of the variability in the endogenous constructs.

The single arrowhead line shows the structural influence from end construct (*j*) to head construct (*i*). The head construct is a dependent or effect construct;

> *Example   Continued*
> the end construct is the causal construct. An exogenous construct (or variable) in a structural equation model is one which has no single headed arrowhead lines entering it, thus, it is not caused by any other construct in the model; it is an independent construct. An endogenous construct (or variable) is one which has at least one single headed arrowhead line entering it, thus, it is caused by one (or more) construct(s) in the model. The β_{ij} coefficients are the structural (regression) coefficients explaining relationships from exogenous constructs to endogenous constructs which are mediated (indirect) by other endogenous constructs. The γ_{ij} coefficients explain relationships which are direct from exogenous constructs to endogenous constructs.
>
> The implication of a specific coefficient is the expected change in explained construct caused by one unit change of the corresponding explanatory constructs whilst holding all the other explanatory constructs and error terms constant. The double-arrowhead curve shows that there is hypothesised covarying relationship between the two exogenous constructs but the cause underlying them will not be identified in this research model. ϕ_{ij} is the correlation coefficient.

In path analysis, the underlying assumptions are:

(1) The exogenous and endogenous constructs are measured with no or negligible error and have an expected value of 0 $[E(X) = E(Y) = 0]$.
(2) The structural linkage from exogenous to endogenous constructs is linear and additive (Bobko 1990) – the fundamental assumption of linearity in ordinary regression.
(3) The error terms in ζ (a) have a mean of 0 $[E(\zeta) = 0]$ and a constant variance across observations; (b) are independent, i.e. uncorrelated across observations; (c) are uncorrelated with the exogenous constructs; and (d) are uncorrelated across equations, i.e. the variance/covariance matrix of ζ is one diagonal matrix.

A series of structural equations in terms of endogenous constructs implied by the example can be written in the form of various explaining (endogenous and/or exogenous) constructs. For instance, the equations are:

$$\eta_1 = \gamma_{12}\xi_2 + \gamma_{13}\xi_3 + \gamma_{14}\xi_4 + \zeta_1$$

$$\eta_5 = \beta_{53}\xi_3 + \beta_{54}\xi_4 + \gamma_{51}\xi_1 + \zeta_5$$

Structural equations modelling is a statistical analysis that tests the viability of alternative causal explanations of variables that correlate with one another. Since we cannot infer causality from correlation, it is important to remember that structural equations modelling provides information regarding the *plausibility* of causal hypotheses BUT not the conclusions about causality.

SEM mathematically compares the correlation matrix implied by a particular hypothesised model to the real correlation matrix based on the data collected. The analysis examines the *fit* of the correlation matrix from the hypothesised model and that from the real data. SEM provides a *goodness of fit index* which indicates how well the hypothesised model fits the data. SEM becomes complex when multiple measures are used to improve the measurement of each construct – latent variable. When single measures of each construct are used, researchers sometimes call it *path analysis*.

Example
Liu and Fang (2006) develop a power-based model of project leadership using structural equations modelling. The resulting model shows that the motivational function of good leadership operates through managing power gaps by means of power-sharing and power-amassing. The project manager's inherent personal traits and credentials are critical to his/her power exercising so as to motivate members to secure management effectiveness.

Analytic hierarchy process

The analytic hierarchy process (AHP) was developed and documented primarily by Thomas Saaty (1980, 1982). The strengths of the AHP method lie in its (1) ability to decompose a complex decision problem into a hierarchy of subproblems; (2) versatility and power in structuring and analysing complex decision problems; and (3) simplicity and ease of use. However, one major criticism of the AHP is the problem of rank reversal when the introduction of a new alternative reverses the rankings of previously evaluated alternatives (see Belton & Gear 1983, Dyer 1990).

The top level in the hierarchy consists of only one element – the overall objective. Subsequent levels may each have several elements, usually between 5 and 9 (Saaty 1980, p. 28). Once the hierarchy is established, priorities (relative importance weights) must be established for each set of elements at every stage of the hierarchy. Finally, the weighted evaluation of each alternative is obtained by summing the weighted scores (by multiplying the priority weight and the evaluation rating) of all attributes.

Canada (1996) summarises the five stages used in AHP as follows:

(1) Construction of a decision hierarchy of decision elements and identifying decision alternatives.

(2) Determination of the relative importance of attributes.

(3) Determination of the relative weight of each alternative with respect to each next higher level attribute. Priority data are obtained by asking various decision makers to evaluate a set of elements at one hierarchical level in a pairwise fashion regarding their relative importance with respect to an element in the next higher level of the hierarchy. After obtaining the pairwise judgements, the next step is the computation of a vector of priorities (or weighting of the elements in the matrix). In terms of matrix algebra, this consists of calculating the principal vector (eigenvector) of the matrix and then normalising it to sum to 1.0 or 100%.

(4) Determination of indicators of consistency in making pairwise comparisons.

(5) Determination of the overall priority weight (score) of each alternative. The final result is obtained from calculation of the vector of the overall priority weights of alternatives. For all i attributes, weighted evaluation for alternative $k = \Sigma$ (priority weight$_i \times$ evaluation rating$_{ik}$).

Examples

A project procurement selection model is developed by Alhazmi and McCaffer (2000) using AHP.

Mahdi *et al.* (2002) develop a Decision Support System to select contractors using the Delphi method and the results of AHP.

Analysing documents (from texts)

Atkinson and Coffey (1997, p. 47) assert that '... we cannot treat records – however "official" – as firm evidence of what they report'. That sentiment is especially pertinent for more overtly political situations – it has been known for regimes not only to selectively destroy books and historical documents, but also to 're-write history' from a perspective favourable to them and in accordance with their dogma. Indeed, even construction project records represent the outcome of negotiations (in most cases), e.g. valuations of variations and delay claims, EOT awards and final accounts.

People produce, use and interpret documents. Hence, judgements, perspectives, power etc. are relevant to documents – what is included as written and what is understood from what is read. In turn, such factors impact on (whether and) how and what an organisation learns. Orton (1997) notes that '... historical documents ... are non-reactive'. However, in follow-up research, often through interviews, to seek explanation or to verify a researcher's interpretation, Orton observes '... when researchers ask a manager why he or she decided to change an organization structure, the researcher and the manager collaborate to impose retrospectively a reality which may not have existed ...'. Such *post-hoc rationalisation* is attributed to '... hindsight biases and retrospective rationality ...'.

Most professions, industries, etc. have developed quite distinctive conventions and styles of writing. Including jargon, there are many specialised uses of language, often accompanied by 'shorthand' as well. Thus, the production and use of most documents assumes a degree of familiarity and expertise for the discipline(s) in question for correct production and interpretation – analogous to the dialect(s) of a language.

Where a multiplicity of documents is involved, there is likely to be a formal prescribed hierarchy – as for a construction project's contract documents. This is important to resolve ambiguity, conflict of contents, etc.

In a more generic research sense, Latour and Woolgar (1986) examine scientific documents – which often take on an individual existence independent of the author(s), and in so doing acquire external authority/credibility.

It is always vital to consider the whole of a document and its context, and especially important to appreciate the extent of applicability etc., i.e. the validity of the document and, more particularly, its contents. It is common for statutes (ordinances, etc.) to include a statement of the intent of the legislation at the beginning of the document; in courts of law (and similar institutions), an important consideration for resolution of contractual disputes is to determine the over-riding intent of the parties.

Thus, many researchers (e.g. Rose 1960) believe that phenomena (things) cannot exist separately from the words which are used to describe them – i.e. they exist through the words used. Further, texts have structuring effects to indicate appropriate actions (e.g. the obvious instance of procedures manuals – such as for quality assurance regulation or the procedures described in contracts – often in 'if/then' terms).

Whatever the intent of the producer and whatever the contents of a document, making sense of the document via interpretation is an activity which is unavoidably undertaken by the reader.

Thus, like other data, documents cannot be regarded as 'independent facts' but as items which are subject to a number of subjective aspects – all of which should be taken into account in their use (what we might call 'holistic constructivism').

Conversation analysis

Conversation analysis concerns the 'institutional order of interaction' (Goffman 1955) but it relates also to social ordering in interactions. Thus, conversations are analysed in terms of the structural and content aspects of oral interchanges (i.e. includes sequencing, pauses, gestures, grammar, opening and closing).

Thus, Heritage (1997, p. 162) notes that conversation analysis '... focuses ... on issues of meaning and context in interaction ... by linking both meaning and context to the idea of sequence, ... sequences of actions are a major part of what we mean by context, that the meaning of an action is heavily shaped by the sequence of previous actions from which it emerges, and that social context is a dynamically created thing that is expressed in and through the sequential organization of interaction'. Thus 'talk is context shaped, in which people create (or maintain or renew) a context for the next person's talk'. 'The assumption is that it is fundamentally through interaction that context is built, invoked and managed, and that it is through interaction that institutional imperatives originating from outside the interaction are evidenced and made real and enforceable for the participants' (Heritage 1997, p. 163).

In particular, Heritage advances six aspects to examine in conversation analysis to reveal the institutionality of interaction.

(1) Turn-taking (who speaks when, and how the changes between speakers occur).
(2) Overall structure of the conversation (constructing a 'map' of the conversation, regarding the main phases/sections, such as 'opening', 'issue', 'introduction', 'response', 'discussion', 'closing'). The structure emerges from the conversation – analogous to grounded theory – and should not be a preconceived structure imposed by the researcher/analyst.
(3) Organisation of sequence (examination of how ideas are initiated and followed up and how others are excluded).
(4) Turn-design: Drew and Heritage (1992) identify two components of turn-design: (a) the action which an individual desires to accomplish by taking a turn in the conversation – the question/message to be delivered; (b) selection of how the question/message is delivered.
(5) Lexical choice (selection of terms etc.): Drew and Heritage (1992) note the context sensitivity of descriptions/terms in that people tend to select terms which fit particular settings or roles.
(6) Interaction asymmetries, which may occur in a variety of contexts: professional–lay conversations; knowhow – for one participant the situation is routine (job) whilst for the other it is personal; knowledge (epistemological caution) – an expert with particular knowledge may be constrained in making very definite statements. Further, experts may exert 'superiority' by their choice of words (lexical choice). Access to knowledge concerns what is known and how it is known and may also depend on role

in an encounter: a participant may wish to avoid giving the impression of being 'nosey'.

Heritage identifies two major branches of conversation analysis. The first 'examines the social institution of interaction as an entity in its own right, the second studies the management of social institutions (such as corporations, ..., medicine etc.) in interaction'. The former analysis concerns conversations between professions or professionals and clients.

Conversation analysis is stringent in requirements of empirical grounding. Other types of discourse analysis and social constructionism tend to emphasise that language may be interpreted in various ways to yield different meanings and so, those approaches pay much attention to the role of the researcher in determining the description of the use of language and, hence, the meanings.

Sacks *et al.* (1974) note that conversation analysis concerns the ways in which social realities and relationships are constituted through persons' talk-in-interaction. The interpretations and requisite methods emerge from the structure and processes of conversations. Hence, conversation analysis must be very local in approach with great attention to detail. Beyond the words employed, the analysis considers sequence, verbal tones, orientation to others, turn-taking etc. to yield a holistic analysis. Transcripts may be analysed through context-sensitive and context-free perspectives.

Discourse analyses

Discourse concerns consideration of statements about a subject at some length, whether oral or written. It varies from casual conversation etc. due to likely, assumed purpose. Discourse analysis has a number of origins (and hence, varied traditions of approach): cognitive psychology, linguistics, sociolinguistics, poststructuralism social psychology and communications, which are considered variously, individually or in combination. A particular branch of discourse analysis relates to the work of Michel Foucault and focuses on how discourse comes to constitute subjects/objects; further, it helps to identify related practitioners as persons with knowledge, authority and (hence, often) power.

The sociological and communications approach to discourse analysis '... emphasises the way versions of the world, ... events and inner psychological worlds are produced in discourse ... this leads to concern with participants' constructions and how they are accomplished and undermined; and ... to a recognition of the constructed and contingent nature of researchers' own versions of the world' (Potter 1997, p. 146).

Thus, discourse analyses analyse talk and texts as manifestations of social practices. Typically, study employs transcripts of talks, speeches, interviews etc. and similarly derived documents for analysis. The analysis is mainly qualitative, rather than the quantitatively oriented analyses of coding and counting constituents of the discourses. Potter (1997, pp. 147–148) considers discourse analysts to employ craft skills and develop an analytic mentality. Thus, 'norms are oriented to; that is they are not templates for action, but provide a way of interpreting deviations'. The analysis tends to be acceptable provided it demonstrates deviations clearly and accurately – to teach a person how to swim involves more than explaining technicalities, but the deviance of non-swimming can be identified readily (as when somebody sinks).

Foucauldian discourse analysis involves '. . . configurations of assumptions, categories, logics, claims and modes of articulation' (Miller 1997, p. 32). Discourse concerns particularities afforded to communication according to the context (e.g. legal) to assist the organisation and sense-making of practical aspects of life. Thus, Miller (1997) notes that 'Foucauldian discourse studies involve treating the data as experiences of culturally standardized discourses, that are associated with the particular social settings.'

Hence, individual discourses are particular to circumstances/context and so, involve particular meanings (at the same time tending to exclude alternative interpretations) which may be manifested in jargon and represent a situation of indexicality.

A further aspect of discourses concerns the possession and structures of knowledge and power. For many participants in the construction industry, it is important to appreciate a variety of discourses and, hence, have knowledge of diverse subject areas (e.g. engineering, architecture, contracts) in order to exercise and enjoy appropriate power in suitable ways as well as the requirements of exercising professional/industrial roles.

Example
Fernie *et al.* (2006) analyse discourses in construction publications to provide a critical perspective on the forces for change in the industry, which have been widely promoted and pursued (largely, dogmatically) for some considerable time.

Social network analysis

Social network theory is evolving and spans across a variety of social science disciplines but tends to be focussed in sociology. The method involves collecting data to produce a diagram (map, network) which depicts individual actors (persons, organisations) as nodes, and the relationships between them as ties. The networks can be used to indicate the social capital of each actor.

Networks vary in many ways – type of actors, size, structure, natures of ties, etc. By concentrating on the network rather than the individual actors, social network theory focuses on relationships between the actors as the important element in the operation (efficiency, effectiveness) of the network. It seems that more open networks in which actors are connected by numerous, but fairly weak, ties enable new ideas to be introduced more readily. For organisations, network theory is useful in plotting and analysing informal systems of communications and relationships, including bridging relationships between persons employed by different companies – important in the TMOs of construction projects. Such studies are germane to determination of power structures as actors close to the centres of social networks are able to exercise social power and influence (often more important than official job title).

Marsden (1990) provides an extensive review of social network theory, including raising several fundamental questions which any researcher employing social networks should address. Those questions include:

- definition of what constitutes a social relationship
- whether actual social relations are to be measured or those perceived by the actors involved (cognitive networks)
- the time frame for the relationships, including when relationships begin, change and end
- where the boundaries of the relationships lie – realist approach is based on the actors' perceptions; nominalist approach is based on the observer's view. Commonly, boundaries are delineated by membership criteria for formal organisations, by social relations as identified in snowball sampling, or by participation in events.

A particular issue concerning the construction of accurate representations of social networks is that people seem to have good recall of typical interactions but are much less accurate in reporting on activities undertaken within specific time periods.

Analysis of social networks is accomplished through a variety of metrics. The size of the network is the number of direct ties involving individual units (also measures integration, popularity and range). Network density is the mean strength of connections between the units in a network; Rowley (1997) defines density as '. . . the ratio of the number of relationships that exist in the network . . . compared with the total number of possible ties if each network member were tied to every other member'. Increasing density of networks improves efficiency of communication and so, norms are readily diffused amongst the units which, then, share behaviours and expectations. Dense networks also promote the formation of coalitions between units.

Centrality measures the relative positions of units in a network; centralisation measures the variability in centrality amongst units. Those measures relate to the power secured by units due to the structure of the network. Rowley (1997) notes the three types of centrality examined by Brass and Burkhardt (1993) as degree centrality (number of direct ties to other units), closeness centrality (independent access to others – sum of the lengths of the shortest paths from a particular unit to linked others), and betweenness centrality (control over other units; frequency with which a unit lies on the shortest paths between other units).

Strengths of ties between units are measured in various ways to reflect the constructs which are important to the purposes of the analysis. Range of a network measures the extent to which a particular unit in the network is linked to diverse other units as measured by network size or network density (inverse measure).

Example

Loosemore (1998) uses social network analysis to facilitate investigation of the causes, processes and consequences of crises in the industry; Pryke (2004) employs social network analysis in application to the concept of construction projects as coalitions of actors.

Multi-level research

A variety of theories in management, economics etc. recognise the impact of the environment on a system; such situational or contingency theories address both endogenous and exogenous changes in their effects on the system operation (and output). Further, directions of impact may be bottom-up as well as top-down and, of course, horizontal.

To deal with such complexity, multi-level research is necessary. Consistency in data collection is essential with maximum 'objectivity' to avoid the problem of differences being due only to different perspectives etc. of respondents at the various levels examined. In investigations such as system mapping in an organisation (e.g. the process, system, followed to produce and submit a tender), the outcome is likely to depend on who is asked. (How are the data collected – each (level of) respondent, director, chief estimator, estimator, etc., is likely to yield a different map of the system due to their differences in knowledge and perception; observations of the operation of the system by the researcher may produce yet another different map.)

The data collection and analysis may proceed via synthesis (progressively adding variables until an adequate model is obtained at each level of individual, groups, etc.) or disaggregation (the progressive splitting of a complex process/model into components and discarding the insignificant variables, again at progressive levels – firm, work groups, individuals, etc.).

Although, ideally, relationships between only two (one independent and one dependent) variables are researched, this is rarely the situation, even in a laboratory experiment, other variables are present and must be taken into account and dealt with in the research design and execution.

If the two variables under study are A and B, the relationship between them may be affected by other variables in four main ways. An intervening variable (X) is where A affects X and then X affects B; by holding X constant and calculating the partial correlation coefficient between A and B, the impact of X is removed and, hence, the true relationship between A and B is revealed. In a chain relationship, A affects X, X affects B and B affects A (also a circular relationship). Theory is the best basis for interpretation of results. Where a confounding, or antecedent, variable is present, X affects both A and B; as for an intervening variable, X is held constant and the partial correlation between A and B is determined. A moderating variable affects the difference between A and B; that X is present as a moderating variable would be revealed in the regression equation to predict B by a significant impact of the XA term.

Meta-analysis

As research studies proliferate, it becomes interesting to consider similarities and differences between the results. Thus, meta-analysis has been devised to qualitatively integrate different studies on the same topic to study the variations in results and to be able to predict for a broader population.

Meta-analysis is a statistical procedure used to analyse and integrate the results of many individual studies on a single topic (Cooper 1990). Meta-analysis examines every study that has been conducted on a particular topic to assess the relationship between those variables that constitute the focus of the research.

> *Example*
> Eagly and Johnson (1990) reviewed previous research on gender differences in leadership style using meta-analysis and found male and female differences in laboratory studies but not in actual business organisations, i.e. in actual business organisations, none of the variability in managers' leadership styles was variance due to gender.

Although details of the integrating method are beyond the scope of this book and revolve around correlation and analysis of variables (using computing), certain issues are noteworthy. The topic must be identified precisely, preferably with a narrow scope, to identify the studies to be included. Integration issues concern coding of the research elements – explicit, comprehensive and mutually exclusive categories are vital – and the means by which the results of the various studies will be combined (usually into a single measure).

Longitudinal research

Much research is, of necessity, cross-sectional; resource constraints, notably time, dictate that data can be collected at one time (instant) only. Such research design means that the establishment of causality is more problematic and so heavy reliance is placed on existing theory/knowledge. If research can be carried out over longer periods, using two or more occasions of data collection (or continuous data collection over a significant period, e.g. video, time-lapse photography, diaries), establishing causality can be done more readily.

Longitudinal research, really data collection, uses either discrete time design, where cross-sectional data are collected on two or more occasions, or continuous time design, where data are collected 'continuously' over a period. Choices of time intervals or periods are important to ensure the data capture the full range of effects and are not 'selective'.

Drenth (1988) notes two forms of discrete time design. In *cross lagged panel design*, the same group/sample of respondents are questioned on at least two occasions (t_1, t_2 etc.), separated by a time interval. Such design is used to examine causal relationships between two variables (A, B) by asking the same question on each occasion to detect changes of opinion etc.; the approach helps to understand any causal relationship which may not be apparent from theory (alone). The score on B at t_1 is influenced by the score on A at t_1; also at t_2. Further, at t_2 the score on B is influenced by the scores on both A and B at t_1. A similar situation applies to A then, if the effect of (At_1 on Bt_2) > (Bt_1 on At_2) it is apparent that A causes B.

The other discrete time design is difference scores, in which the magnitudes and directions of changes in scores on the variables over the time interval are examined. However, the changes in scores are of doubtful reliability (compared with the actual scores obtained), are likely to be subject to 'ceiling' (and 'floor') effects and are also likely to be subject to regression effects.

Generally, the discrete time approach is problematic due to the lack of predictability and generalisability of the influences of different time periods (intervals) on change in the scores of the variables in both magnitude and direction.

In continuous time design, data are collected continuously over a period so that the changes in variables are measured as they occur. A common method is to ask respondents questions about the past, although, especially for more distant events, qualitative data (change of 'state' of a person, e.g. unmarried to married) are more likely to be reliable than quantitative data (level of happiness in the two 'states'). Thus, triangulation may be helpful – such as use of diaries or discrete time data collection to supplement the retrospective questioning.

Summary

This chapter has considered several techniques for the analysis of data. It is useful to aim to maintain simplicity; understanding what analyses are being undertaken and why, and their validity, is *paramount*. Usually, it is helpful to plot the raw data to gain a first impression of any patterns. We have examined the basics of regression and correlation, time series and index numbers. It is important to use only those tests and techniques which are appropriate to the data, so awareness of the nature of the data collected is vital. Often, results in the form of hierarchies will be obtained – in such instances, rank correlations may be useful. In cases of complexity, it is likely to be a good idea to consult the tutor, supervisor or/and a statistician.

References

Akintoye, A. (2000) Analysis of factors influencing project cost estimating practice, *Construction Management and Economics*, **18**(1), 77–89.

Aldenderfer, M.E. & Blashfield, R.K. (1984) *Cluster Analysis*, Sage, Beverly Hills, CA.

Alhazmi, T. & McCaffer, R. (2000) Project procurement system selection model, *Journal of Construction Engineering and Management*, **126**, 176–184.

Atkinson, P. & Coffey, A. (1997) Analysing documentary realities. In: *Qualitative Research: Theory, Method and Practice* (ed. D. Silverman), pp. 99–112, Sage, London.

Belton, V. & Gear, T. (1983) On a short-coming of Saaty's method of analytic hierarchies, *Omega*, **11**(3), 228–230.

Bobko, P. (1990) Multivariate correlational analysis. In: *Handbook of Industrial and Organizational Psychology* (eds M.D. Dunnette & L.M. Hough), Consulting Psychology Press, Palo Alto, CA.

Bollen, K.A. (1989) *Structural Equations with Latent Variables*, Wiley, New York.

Brass, D.J. & Burkhardt, M.E. (1993) Potential power and power use: an investigation of structure and behavior, *Academy of Management Journal*, **36**(3), 441–470.

Canada, J.R. (1996) *Capital Investment Analysis for Engineering and Management*, Prentice-Hall, Englewood Cliffs, NJ.

Charmaz, K. (1995) Between positivism and postmodernism: implications for methods, *Studies in Symbolic Interaction*, **17**, 42–72.

Comrey, A.L. & Lee, H.B. (1992) *A First Course in Factor Analysis*, Lawrence Earlbaum Associates, Hillsdale, NJ.

Cooper, H. (1990) Meta-analysis and the integrative research review. In: *Research Methods in Personality and Social Psychology* (eds D. Hendrick & M.S. Clark), pp. 142–163, Sage, Newbury Park, CA.

Dillon, W.R. & Goldstein, M. (1984) *Multivariate Analysis: Methods and Applications*, Wiley, New York.

Drenth, P.J.D. (1998) Research in work and organizational psychology: principles and methods. In: *Handbook of Work and Organizational Psychology*, Vol. 1, 2nd edn, (eds P.J.D. Drenth, H. Thierry & C.J. de Wolff), pp. 11–46, Psychology Press, Hove.

Drew, D. & Skitmore, R.M. (1997) The effect of contract type and size on competitiveness in bidding, *Construction Management and Economics*, **15**(5), 469–489.

Drew, P. & Heritage, J. (eds) (1992) *Talk at Work: Interaction in Institutional Settings*, Cambridge University Press, Cambridge.

Dyer, J.S. (1990) Remarks on the analytic hierarchy process, *Management Science*, **36**(3), 249–258.

Eagly, A.H. & Johnson, B.T. (1990) Gender and leadership style, *Psychological Bulletin*, **108**, 233–256.

Edwards, D.J., Holt, G.D. & Harris, F.C. (2000) A comparative analysis between the multilayer perceptron 'Neural Network' and multiple regression analysis for predicting construction plant maintenance costs, *Journal of Quality in Maintenance Engineering*, **6**(1), 45–60.

Fang, Z.Y. (2002) *Behavioural Analysis of Project Team Performance in China*, Unpublished PhD thesis, Department of Real Estate and Construction, University of Hong Kong.

Fernie, S., Leiringer, R. & Thorpe, T. (2006) Change in construction: a critical perspective, *Building Research and Informaiton*, **34**(2), 91–103.

Forman, E.H., Saaty, T.L., Selly, M.A. & Waldron, R. (1983) *Expert Choice*, Decision Support Software, McLean, VA.

Geddes, P. (1968) *Cities in Evolution*, Benn, London.

Goffman, E. (1955) Facework, *Psychiatry*, **18**, 213–231.

Heritage, J. (1997) Conversation analysis and institutional talk: analysing data. In: *Qualitative Research: Theory, Method and Practice* (ed. D. Silverman), pp. 161–182, Sage, London.

Kaiser, H.F. (1958) The varimaxcriterion for analytic rotation in factor analysis, *Psychometrika*, **23**, 187–200.

Ketchen, D.J. & Shook, C.L. (1996) The application of cluster analysis in strategic management research: an analysis and critique, *Strategic Management Journal*, **17**, 441–458.

Kleinbaum, D.G., Kupper, L.L. & Muller, K.E. (1988) *Applied Regression Analysis and Other Multivariable Methods*, PWS-Kent, Boston, MA.

Labovitz, S. (1970) The assignment of numbers to rank order categories, *American Sociological Review*, **35**, 515–524.

Latour, B. & Woolgar, S. (1986) *Laboratory Life: The Construction of Scientific Facts*, Princeton University Press, Princeton, NJ.

Levin, R.I. & Rubin, D.S. (1990) *Statistics for Management*, 5th edn, Prentice-Hall, Englewood Cliffs, NJ.

Li, C.C. (1975) *Path Analysis: A Primer*, Boxwood Press, Pacific Grove, CA.

Lingoes, J.C. (1968) The multivariate analysis of qualitative data, *Multivariate Behavioural Research*, **3**, 61–64.

Liu, A.M.M. & Fang, Z. (2006) A power-based leadership approach to project management, *Construction Management and Economics*, **24**(5), 497–508.

Loosemore, M. (1998) Social network analysis using a quantitative tool within an interpretative context to explore the management of construction crises, *Engineering, Construction and Architectural Management*, **5**(4), 315–326.

Mahdi, I.M., Riley, M.J., Fereig, S.M. & Alex, A.P. (2002) A multi-criteria approach to contractor selection, *Engineering, Construction and Architectural Management*, **9**(1), 29–37.

Malhotra, N.K. & Birks, D.F. (1999) *Marketing Research. An Applied Approach*, Prentice-Hall, London.

Marsden, P.V. (1990) Network data and measurement, *Annual Review of Sociology*, **16**, 435–463.

Miller, G. (1997) Systems and solutions: the discourses of brief therapy, *Family Therapy*, **19**(March), 5–22.

Moser, C.A. & Kalton, G. (1971) *Survey Methods in Social Investigation*, Gower, London.

Moses, L.E. (1986) *Think and Explain with Statistics*, Addison Wesley, Reading, MA.

Mueller, R.O. (1996) *Basic Principles of Structural Equation Modeling. An Introduction to LISREL and EQS*, Springer Verlag, New York.

Nie, N.H., Hull, C.H., Jenkins, J.G., Steinbrenner, K. & Brent, V.H. (1975) *SPSS: Statistical Package for the Social Sciences*, McGraw-Hill, New York.

Norusis, M.J. (1992) *SPSS/PC+ Base System User's Guide Version 5.0*, SPSS, Chicago.

Nunnally, J.C. (1978) *Psychometric Theory*, McGraw-Hill, New York.

Orton, J.D. (1997) From inductive to iterative grounded theory: zipping the gap between process theory and process data, *Scandinavian Journal of Management*, **13**(1), 419–438.

Parry-Lewis, J. (1965) *Building Cycles and Britain's Growth*, Macmillan, London.

Pindyck, R.S. & Rubinfeld, D.L. (1981) *Econometric Models and Economic Forecasts*, 2nd edn, McGraw-Hill, New York.

Popper, K. (1989) *Conjectures and Refutations: The Growth of Scientific Knowledge*, Routledge & Kegan Paul, London.

Potter, J. (1997) Discourse analysis as a way of analysing naturally occurring talk. In: *Qualitative Research: Theory, Method and Practice* (ed. D. Silverman), pp. 144–160, Sage, London.

Pryke, S.D. (2004) Analysing construction project coalitions: exploring the applications of social network analysis, *Construction Management and Economics*, **22**(8), 787–797.

Rose, E. (1960) The English record of a natural sociology, *American Sociological Review*, **XXV**(April), 193–208.

Rosenthal, R. & Rosnow, R.L. (1991) *Essentials of Behavioral Research. Methods and Data Analysis*, McGraw-Hill, New York.

Rowley, T.J. (1997) Moving beyond dyadic ties: a network theory of stakeholder influences, *The Academy of Management Review*, **22**(4), 887–910.

Rummel, R.J. (1970) *Applied Factor Analysis*, Northwestern University Press, Evanston, IL.

Saaty, T.L. (1980) *The Analytic Hierarchy Process*, McGraw-Hill, New York.

Saaty, T.L. (1982) *Decision Making for Leaders*, Wadsworth Publishing, Belmont, CA.

Sabherwal, R. & Robey, D. (1995) An empirical taxonomy of implementation processes based on sequences of events in information system development. In: *Longitudinal Field Research Methods. Studying Processes of Organizational Change* (eds G.P. Huber & A.H. Van de Ven), pp. 228–266, Sage, Thousand Oaks, CA.

Sacks, H., Schegloff, E.A. & Jefferson, G. (1974) A simplest semantics for the organization of turn-taking for conversation, *Language*, **50**, 696–735.

Scheiner, S.M., Mitchell, R.J. Callahan, H.S. (2000) Using path analysis to measure natural selection, *Journal of Evolutionary Biology*, **13**(3), 423–433.

SPSS (1997) *SPSS Base 7.5: Applications Guide*, SPSS, Chicago.

Ulrich, D. & McKelvey, B. (1990) General organizational classification: an empirical test using the United States and Japanese electronic industry, *Orgainzation Science*, **1**, 99–118.

Ward, J.H. Jr (1963) Hierarchical grouping to optimize an objective function, *Journal of the American Statistical Association*, **58**, 236–244.

Wilson, M. (1995) Structuring qualitative data: multidimensional scalogram analysis. In: *Research Methods in Psychology* (eds G.M. Breakwell, S. Hammond & C. Fife-Shaw), pp. 259–275, Sage, London.

Yeomans, K.A. (1968) *Statistics for the Social Scientist: 2: Applied Statistics*, Penguin, Harmondsworth.

Chapter 8

Ethics in Research

The objectives of this chapter are to:

- introduce the concepts of morals and of ethics

- examine applications of *research ethics* in literature review and data collection, data use and disposal

- introduce principles and practices concerning data analysis, intellectual property and data protection.

The concepts of morals and ethics

One of the important concerns in research is that ethical issues are far more extensive than is usually recognised. Indeed, one may suggest that there are ethical considerations relating to everything people do and consider doing. In research, the attention which is given to ethics recently has had its greatest focus on collection of data from human 'subjects' and the storage, use and disposal of those data. Whilst that is an important area, perhaps most obviously in the social sciences (and, hence, management), it seems to represent an extension of legal attention to data protection and intellectual property. Thus, it is important to be aware of the many other aspects of moral and ethical concerns for research, including use of the work of other people, confidentiality and integrity in collecting data, analysing data and reporting and disseminating results and findings.

The *Oxford English Dictionary* (2007) defines 'moral' as:

- Of or relating to human character or behaviour considered as good or bad; of or relating to the distinction between right and wrong, or good and evil, in relation to the actions, desires, or character of responsible human beings; ethical.

- Of an action: having the property of being right or wrong, or good or evil; voluntary or deliberate and therefore open to ethical appraisal. Of a person, etc.: capable of moral action; able to choose between right and wrong, or good and evil.
- Of knowledge, an opinion, etc.: relating to the nature and application of the distinction between right and wrong, or good and evil.

Thus, morals comprise the fundamental beliefs of people over what is right and what is wrong and so, underpin behaviour – human actions and interactions.

The *Oxford English Dictionary* (2007) defines 'ethics' as:

- The science of morals; the department of study concerned with the principles of human duty.
- The moral principles or system of a particular leader or school of thought.
- The moral principles by which a person is guided.
- The rules of conduct recognised in certain associations or departments of human life.
- In wider sense: The whole field of moral science, including besides Ethics properly so called, the science of law whether civil, political or international.

Hinman (1997) distinguishes morals and ethics by regarding morals as first-order beliefs, and practices about what is good and what is bad which guide behaviour and ethics as second order, reflective consideration of moral beliefs and practices. Snell (1995, p. 155) notes that moral ethos, which may be a strong force to encourage or constrain action, is, '... the social climate predisposing members of an organization toward adopting and enacting some particular ethical standards and deflecting them away from others.'

Because research involves the furtherance of (human) knowledge, the requirement of ethical integrity is paramount – and, perhaps illustrated best by reflecting on medical research, such as the safety of medicines. Whilst, at the more extreme level, immoral or unethical behaviour is driven by a desire to defraud others (often, for individual gain), a more widespread possibility is failing to exercise adequate care – consciously or, even, unconsciously. It is about ensuring that the results are as robust as possible and declaring what has been done to ensure they are robust – it's not just do not fabricate the data, don't 'fiddle the results'. Thus, it is highly desirable to include express measurements of confidence and significance whenever possible.

Before addressing ethics in research, a primary problem is that there is a diversity of basic, philosophical perspectives of ethics. Those perspectives yield three primary paradigms for ethical behavioural analysis (see Leary 2004, p. 345).

- In deontology, a universal moral code is held to apply.
- In scepticism (alternatively, relativism; subjectivism) ethical rules are arbitrary and relative to culture and to time; that view is extended in ethical egoism where ethics becomes a matter of the conscience of the individual such that the individual is the arbiter of ethics (what is right and what is wrong). Thus, egoism concerns pursuit of self-interest and so, can be related to common business criteria (notably, profit maximisation).
- Teleology constitutes a utilitarian approach (utilitarianism; consequentialism) where ethics are dependent upon the (anticipated) consequences – that suggests a cost–benefit view, perhaps invoking the judgmental criterion of 'the greatest good for the greatest number' which, itself, is likely to necessitate subjectively determined weightings and value judgements.

Other forms of ethics include eudaemonistic ethics in which the values to facilitate greatest self-actualisation are adopted. Virtue ethics focuses on the merits of facets of character and relates to deontology/eudaemonism. Pursuit of pleasure (hedonism) is often contrasted with religious ethics (again, related to deontology and to egoism).

Given the variety of ethical perspectives, it is hardly surprising that dilemmas are quite common. Hence, individuals are forced to choose; such choices are context bound – by culture, laws, codes of conduct and so on. Although it is apparent that, at the primary (basic) level, a set of universal 'goods' exist, at the secondary (culturally shaped definitions and codes of conduct) and tertiary (specific codes of behaviour) levels, notable differences apply.

The organisational ethics model in Fig. 8.1 adopts a systems approach to show transformation of behaviour from (ethical) goals to outcomes as well as identifying endogenous and exogenous variables from a behavioural control perspective. The goals, behaviour, performance, outcome (GBPO) (Liu 1999) cycle is employed as the basic cognitive logic for the model. Three levels of ethical influence, namely, individual, local (internal groups/teams, i.e. the organisation members) and cosmopolitan (external/societal) are postulated (Victor & Cullen 1988). The local level is particularly influential on individuals as it is their immediate environment and it is on the local level in the organisation context that the research using the model is focused.

The individual, local and cosmopolitan levels of ethical influence are employed to facilitate examination of impacts between the three levels, as well as considering internal–external orientation. The concept of levels of ethical development is also in line with Kohlberg's (1981) idea of an individual's progression of ethical reasoning from one stage to another. The Kohlbergian

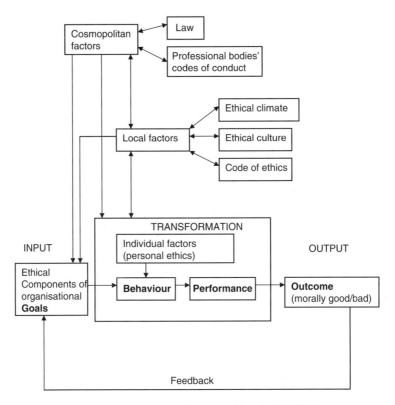

Fig. 8.1   Model of organisational ethics (*source:* Liu *et al.* (2004)).

model is adapted by Snell (1995) to give direct applicability to managers and professionals whereby one progresses from obedience to authority, to conformance to rules and codes, to continually questioning one's own actions and principles.

Consistency is an issue in determining whether an action is ethical – which requires the establishment of 'standards'. However, notably in the context of research, change is 'a constant' rendering the establishment and application of standards more problematic – in a context of change, the only standards which are useful relate to 'process' rather than 'product' (so, 'use the most robust test', rather than, 'use test θ').

Research ethics

A particular issue which permeates ethics, including ethics in research, is that, by its very nature, ethics is a subject which is culturally bound. It is not only that a diversity of ethical theories exists but, further, that ethics means different things in different places and so, leads to differences in common, acceptable practices which may change over time. It is essential for such variabilities to be taken into consideration in formulating and conducting research (e.g. practices of paying gratuities, if any; is a gratuity really optional; when does a gratuity become a bribe?).

The ESRC (Economics and Social Research Council) (2007, p. 7) in the UK defines research as '. . . any form of disciplined inquiry that aims to contribute to a body of knowledge or theory' and continues that ' "Research Ethics" refers to the moral principles guiding research, from its inception through to its completion and publication of results and beyond.' The ESRC expects that six key principles of research ethics will be addressed whenever possible (in grant applications which it considers and in the execution of research projects) (p. 1):

- 'Research should be designed, reviewed and undertaken to ensure integrity and quality.
- Research staff and subjects must be informed fully about the purpose, methods and intended possible uses of the research, what their participation in the research entails and what risks, if any, are involved . . .
- The confidentiality of information supplied by research subjects and the anonymity of respondents must be respected.
- Research participants must participate in a voluntary way, free from any coercion.
- Harm to research participants must be avoided.
- The independence of research must be clear, and any conflicts of interest or partiality must be explicit.'

The application of the general principles to the conduct of research must address the practicalities of carrying out research with people as subjects. Within limits, it is apparent that the teleological approach prevails. That requires the exercising of value judgements by all concerned; in practice it means that research organisations set up research ethics committees, populated by experts from relevant and diverse disciplines who are experienced researchers themselves, to scrutinise proposals for compliance with ethical requirements and monitoring research processes and outputs. Further, most research institutions (notably, universities and major research funding organisations) produce their own code of ethical conduct in research to guide their researchers in producing proposals, conducting studies and disseminating outputs. The Research Ethics Framework (REF) of the UK's Economic and Social Research Council is a leading example (ESRC 2007).

The Belmont Report (1979) in the USA, which has been instrumental in setting standards for ethical practices in research, adopts three basic ethical principles:

- Respect for persons
- Beneficence
- Justice.

Respect for persons requires that individuals are regarded and treated as autonomous agents able to exercise free self-will in making decisions (about whether to participate in research). This requires that persons participate in research voluntarily and have reached that decision after evaluating adequate, accurate information provided by the researcher. However, certain persons have diminished autonomy (e.g. children, persons of diminished mental capacity) and so, must be afforded adequate protection, perhaps through assistance of other, responsible individuals to assist them in deciding whether to participate.

Beneficence concerns respecting person's decisions, protecting them from harm (due to their participation in the research) and endeavouring to ensure their well-being. Here, an ethical dilemma may arise regarding which of the following two ethical principles is adopted – 'do not harm' (absolute, deontological) or 'maximise possible benefits and minimise possible harms' (teleological, utilitarian); the latter should prevent any research which is likely to result in net harm by allowing only research which is anticipated to yield net benefit to proceed. Clearly, as soon as an absolute principle is 'relaxed' value judgements may take over to impact the outcome/decision.

Justice concerns distribution of benefits and burdens, commonly examined under (one of) five primary formulations – an equal share to each person; to each person, according to their individual need; to each person, according to their individual effort; to each person, according to societal contribution; to each person, according to merit. Again, the formulations reflect different ethical principles. Further, justice relates to the selection of possible participants also and so, should be considered in conjunction with the issues of the research population, sampling and other methods of selecting the sources of data. The maxim should be to select data sources on the basis of their relationship to the subject matter of the research, acknowledging practicalities. It is always, at least, questionable – and so, requiring justification, if persons (as data sources) are selected because of their ease of availability, easy manipulation or some compromised position.

Theory and literature

The first activity of carrying out a research project in which ethics plays an important, but, often, unrecognised, role is in reviewing theory and literature.

It is here that the emphasis is on the (past) research of other people and so, must be acknowledged and reported as such. That requires thorough searching and rigorous record keeping to ensure that accurate and complete citations can be included and applies not only to items which are quoted directly or paraphrased (the references) but also to items which constitute the general informative background (bibliography). Unintentionally omitting a reference is a human mistake but intentionally passing off someone else's work as one's own is plagiarism.

Whilst direct quotations can be checked for accuracy of content easily, there remains the aspect of context – the researcher must ensure that the context is represented accurately too; both content and context are more difficult to verify if passages have been paraphrased and, if included incorrectly, misrepresent the original work.

Accessing theory and literature must address concerns of copyright (part of the array of 'intellectual property'). Internationally and domestically, laws exist to protect the rights of original authors over the reuse of their publications by others, including citation requirements. That raises issues over basic rights to knowledge and use of published information, especially in cases where information is increasingly expensive (as for annual subscriptions to, some, journals) and where payments of royalties are involved.

It seems regrettable that the scientific communities are becoming more commercially competitively oriented and so, increasingly protective and secretive over discoveries and ideas. Assessments of research which become fundamental to career maintenance and advance exacerbate the difficulties both by fostering restriction of access to outputs and encouraging as much and as many publications of outputs as possible – what the consequences are for standards of media items is open to debate!

As researchers, we tend to accept the reliability of publications (and non-published outputs, such as dissertations and theses) almost at 'face value'. Theses, papers in refereed journals, research reports, dissertations tend to be regarded at the foremost level of reliability due to the scrutiny by examiners and referees before publication – that reliability, then, is dependent on the abilities and reliability of the scrutineers – including editors in selecting referees. (That seems somewhat analogous to auditing the annual accounts of a public company – a practical guarantee of reliability but not one which is 'absolutely watertight'.) Other publication forms are regarded as of less reliability with some newspapers towards the bottom of the hierarchy. Materials on the web seem to span the entire spectrum and so, each item must be judged individually. Unfortunately, the de-meriting of certain forms of publication is likely to lead to self-fulfilling prophesies as fallible, pragmatic humans respond to pressures.

Data collection, use and disposal

A preliminary issue is the boundary between research and practice. For most research in construction, that boundary is clear; however, that is not the case for *action research* and may be 'fuzzy' in studies adopting ethnomethodology. Essentially, if any level of research is intended, then the ethical principles and applications should be followed.

An important arena of research ethics is that of collecting, storing, using and disposing of data from, and relating to, human subjects. Much of this arena is not only one of ethics but is covered by legislation also. In UK, the primary legislation is the Data Protection Act, 1998 (see, e.g. www.opsi.gov.uk/acts/acts1998/19980029.htm).

One discipline in which ethics of collecting data had to be addressed was in medicine; the medical professions have maintained a strict code of confidentiality regarding information about patients for many years. Psychology is a highly sensitive area concerning collection of data from humans and the conduct of tests and experiments involving people and other living creatures. In legal contexts, a great deal of data are highly sensitive, and so, must be safeguarded, especially prior to an outcome decision.

Thus, clearly, a considerable diversity of disciplines have practical requirements, in addition to moral and ethical concerns over data which have informed current legislation and practices, for instance, the Belmont Report (1979) advocates three general principles for the ethical conduct of research:

- Informed consent
- Assessment of risks and benefits
- Selection of subjects.

Informed consent (to participate in the research – provide data etc.) is grounded in the provision of adequate and accurate information about the research and requested involvement of the subjects and the potential consequences of involvement to enable them to make an informed decision of whether to participate. That information exchange should be a two-way process in which the potential subjects are free to ask questions of the researcher and, further, are free to withdraw from further participation at any time; the latter two 'freedoms' for the subjects should apply throughout their participation. Occasionally, the nature of a research project is such that informing potential subjects of certain aspects of the project may affect the results significantly and/or impair their validity. In those instances, as much information as possible should be given, including an explanation of the nature of the results sought, and so, the nature of the subjects' participation; a full de-briefing should be provided after the subjects' participation has finished.

Comprehension relates to the content and extent of the information – to facilitate understanding by the intended subjects – to support their decision making; that requires attention to the nature and abilities of the desired subjects. Further, it may be important to obtain (express) permission from other persons so that the intended subjects can participate freely (without fear of 'punishment' for participation etc.). Commonly, that applies in research into organisations and so, the researcher should obtain permission from a high level, authoritative representative of the organisation to approach and gain participation by the intended subjects prior to contacting the subjects themselves.

Participation must be completely voluntary and so, must be free from coercion (a threat of harm) of any kind or level and free from undue influence (an offer of an improper, e.g. excessive sum of money, reward) to secure participation. Further, there must be no unjustifiable pressures on persons to participate – such as could occur from individuals in positions of authority over the potential subjects. Of course, much important research in the social sciences (primarily in USA) has been conducted by academics with (usually, post-graduate) students as the subjects whose willing participation has yielded significant results for advancements of knowledge (see, e.g. Diekmann *et al.*, 1996).

Assessment of the risks and benefits should be examined expressly relating to the anticipated outcomes of the research project and to the execution of the research – in particular, any risks and benefit to potential subjects due to their participation. It may be helpful to array the risks and benefits in tabular form for outcomes and for process(es). The table should denote the hazard (what may cause a risk or benefit event – placing a bet may lead to a financial loss or gain), the probability for each hazard (preferably, noting how the probability has been determined – objectively or subjectively) and the consequent risk outcome (the product of each hazard and its probability) to yield the risk and benefit profile for the research. Determination of the overall, net risk or benefit of the research depends on the inter-relationships of the risks and benefits to denote how they should be combined, assuming the outcomes have been measured in common units! Otherwise, value judgements must be made to make an overall assessment of outcomes which have been measured in different units. Generally, if there is a favourable net benefit, a research project should proceed however, if there is one (or more) hazard with significant probability which would lead to a severe risk (likely harm), such as major physical or mental injury to a subject, then the presence of such a risk may be determined to require the abandonment of that research.

Selection of the subjects for research must be fair both in the procedures adopted for selection and the likely outcomes for the subjects. Thus, random sampling of a population is likely to be considered fair (assuming the population was decided reasonably, fairly with reference to the research topic) and giving each subject $5.00 to use in up to X number of 'gambles' would, provided the 'gambles' are fair and the subjects given a consistent set of appropriate information, lead to fair outcomes, determined by the subjects' decisions and actions.

Usefully, an outline of the research and the basics of the ethical aspects pertaining to subjects are included in an initial letter seeking participation. That letter should include notification of a body to which any complaint relating to the research may be addressed; usually, that is the research ethics committee of the university or other institution which is hosting the research.

Having determined and obtained approval of the data requirements and sources of data, it is an ethical requirement that those are the data collected from the sources identified and that the selection and collection procedures are followed rigorously-collect only and all data identified (and consented) and record accurately. However, given the nature of research, data requirements may change; any changes which are significant to the ethical aspects of the research must be subject to proper scrutiny for ethical compliance.

Part of the ethical considerations regarding the data will be how the data will be used in the research and by whom. Traditionally, original data have been collected, except for macroeconomic and similar studies but, increasingly, there are desires to share data or use data collected by others. Whenever original data are collected, an important concern is whether the data will be available for use beyond the instant project – in other words, what is the boundary regarding use of the data? From a data user's perspective, data may 'decay' in usefulness/validity over time and across geographical and other boundaries but, from a collector's and from a data provider's perspectives, the ethical issues of further uses may be grave concerns.

Boundaries of further uses of data may be stipulated from no further use to making the data freely available. Generally, any possible uses beyond the instant project must be examined for ethical issues and any further availability sanctioned must be agreed by the data providers as part of their informed, willing consent. In practice, the boundaries of further use of the data must be specified; perhaps only aggregated, anonymous data are likely to be available quite freely.

The third category of ethical concerns about data relate to disposal of the data. Possibilities range from total destruction of the data once the research has been completed to making it freely available, perhaps by putting it on a web page. Generally, the form of disposal relates to the sensitivity of the data, the more sensitive, the greater the requirement for early and irretrievable disposal (i.e. destruction). However, it is always appropriate to retain the original data until the particular research has been completed, including any examining or other scrutiny of the output.

Throughout the research, any and all promises of anonymity and of confidentiality must be pursued rigorously. In many countries, such considerations are not only of major ethical importance but are subject to legal controls also.

Data analysis, intellectual property and data protection

Data analysis, results and reporting

Today, it is most usual for data to be input into a computer package for analysis. That inputting is in addition to recording, transcribing, coding, classifying and other manipulations of the data which may be necessary to facilitate analysis. Each and every process relating to data must be done accurately and with integrity; that is vital wherever judgements are involved. Accuracy of inputting and similar 'objective' processes can be ensured in practice by sufficient care and (independent) checking but the same cannot be said of subjective activities, such as selection of categories for the data (although some formal techniques are available – cluster analysis etc.).

It is widely stated that 'techniques are value free'. However, that perspective ignores the fact that the techniques have been developed by people and that their selection and uses are governed by individuals' value judgements. What is important for ethical validity is that any bias, limitations and so on, inherent in the analytic techniques adopted, must be acknowledged but, further, that the techniques selected are the most suitable for the research and that they are used correctly. Use of inappropriate techniques or/and using techniques incorrectly invalidates the results, is unethical if done negligently or with intent to deceive and may be fraudulent (in more extreme instances). (The use of a poker vibrator to compact concrete in test cubes, knowing that it would lead to high strength results, totally invalidates that test, is unethical and fraudulent!)

Thus, much relates to a duty of care, to have the integrity to be informed over what analytic methods are valid and to use them properly – 'if in doubt (about the suitability of a method etc.), find out'.

Reporting the results is where the outcomes of the research (results, conclusions, recommendations) are documented for scrutiny and dissemination. Different audiences may warrant different disclosure in terms of content, details (linguistic) approach and media to be effective. The essence is to disclose accurately and with integrity (express honesty), hence, checking results of ethnographic studies with leading subjects prior to publication is highly desirable to ensure accurate reporting.

> *Example*
> Scientific journal *X* was known to publish papers only if results of experiments had data points exactly on the regression line. A research student had results where, quite normally, data point lay around the regression line. In order for the paper to be published in that journal, the student, with the knowledge of the supervisor, adjusted the data points so that they (as experimental results) lay on the regression line; the paper was accepted (after refereeing) and published. However, the adjusted graph of results appeared in the thesis submitted for a PhD, along with a citation of the published paper. The external examiner questioned the amazing fitting of the data points and the situation was revealed. The student's PhD was failed. (Personal communication; all parties are anonymous for ethical reasons.)

Intellectual property

Intellectual property relates to original creations of a person's mind. Those creations must be both significant in scope, content and originality and must have a tangible manifestation as a product, a publication (of prose, poetry, music, film, etc.), computer software, a design, a trade mark, and so on. The categories of intellectual property which are most germane to research are copyrights and patents. Both copyrights and patents afford the originator legal protection for exclusive use of the items for specified periods. Patents must be registered with a patent office to apply but copyright applies as soon as the work is produced in a permanent form (e.g. on paper). Once a patent is granted, it is owned by the patent holder (the inventor); copyright vests in the producer of the work.

Once a patent or copyright exists, the owner may assign it to another – such as to a production company; payment may be involved. Other producers may produce the patented product (during the period for which the patent is valid) only under license from the patent owner and, usually, in return for a payment. For copyright works, only limited use is permitted (photocopying, quotation of extracts, etc.) and proper citation must be given, certain uses require payment of royalties.

It is important that researchers comply with domestic and international law regarding intellectual property; many research organisations produce guidance leaflets. However, especially for copyright, infringements are very difficult to detect and so, compliance is an important ethical concern. If in any doubt, seek permission to use other's material; most publications carry copyright and note who is the holder.

Data protection

At first glance, the issues of data protection and freedom of information may appear to be diametrically opposed, data protection controls collection, storing and use of data whilst freedom of information endeavours to provide ready access to information and procedures. In many countries, both aspects are governed by legislation which, commonly, treats the two issues complimentarily.

In UK, the primary pieces of legislation are the Data Protection Act (1998) (DPA) and the Freedom of Information Act (2000) (FOIA). The DPA focuses on *personal data* and so, governs the processing of any information which relates to a living individual who can be identified from that information. It legislates for the '... regulation of the processing of information relating to individuals, including the obtaining, holding, use or disclosure of such information ...'. The FOIA concerns '... disclosure of information held by public authorities or by persons providing services for them and to amend the Data Protection Act 1998 ...'.

The DPA adopts eight principles of protection for personal data:

- Data must be obtained for lawful and specified purpose(s)
- Data must not be processed in any manner which is incompatible with the specified purpose(s)
- Data must be adequate and relevant and not excessive for the purpose(s)
- Data must be kept up to date
- Data may be kept for no longer a period than is necessary for the specified purpose(s)
- Data must be processed in accordance with the rights of the subject
- Data must be kept safe from unauthorised access, accidental loss or destruction
- Data must not be transferred to a country outside the European Economic Area unless that country has equivalent levels of protection for personal data.

Under the DPA, relevant data are any which relate to a living person or from which a living person may be identified. The scope concerns all computerised processing of personal data, plus forms of manual records – notably, those which may be converted for electronic processing easily. Anyone who processes personal data must notify the Information Commissioner's Office

(TICO), unless such processing is exempt under the DPA. Generally, data subjects have the right to access personal data relating to them to check its accuracy etc.

While compliance with legal requirements is essential, it does not guarantee that ethical requirements have been met. As legal requirements include administrative procedure necessities (e.g. notification of TICO), it is appropriate to examine ethical and legal compliance in parallel. Philosophically, ethical compliance is likely to be more rigorous, it is vital that the legal administrative requirements are followed – in any case of doubt regarding the requirements, it is wise to consult a specialist.

Summary

This chapter has reviewed the main theories of ethics and their foundations in morals. It has considered applications of ethics at various levels and with particular reference to the research processes. Principles of research ethics and their applications under the ESRC's REF in UK and the Belmont Report (USA) have been discussed. Particular care over data relating to human subjects has been emphasised. Finally, legislation relating to data protection and freedom of information has been noted.

References

The Belmont Report. The National Commission for the Protection of Human Subjects of Biomedical and Behavioral Research (1979) *Ethical Principles and Guidelines for the Protection of Human Subjects of Research*, http://www.hhs.gov/ohrp/humansubjects/guidance/belmont.htm, accessed 11 July 2007.

Data Protection Act (1998) *Elizabeth II*, Chapter 29, TSO, London. (See also http://www.opsi.gov.uk/acts/acts1998/19980029.htm).

Diekmann, K.A. Tenbrunsel, A.E. Shah, P.P. Schroth, H.A. & Bazerman, M.H. (1996) The descriptive and prescriptive use of previous purchase price in negotiations, *Organizational Behavior and Human Decision Processes*, **66**(2), 170–191.

Economics and Social Research Council (ESRC) (2007) *Research Ethics Framework (REF)*, Economics and Social Research Council, Swindon; http://www.esrcsocietytoday.ac.uk/ESRCInfoCentre/Images/ESRC_Re_Ethics_Frame_tcm6-11291.pdf, accessed 12 July 2007.

Freedom of Information Act (2000) *Elizabeth II*, Chapter 36, TSO, London.

Hinman, L.M. (1997) *Ethics: A Pluralistic Approach to Moral Theory*, Harcourt Brace Jovanovich, Orlando.

Kohlberg, L. (1981) *Essays on Moral Development. Vol. 1: The Philosophy of Moral Development*, Harper & Row, San Francisco.

Leary, M.R. (2004) *Introduction to Behavioural Research Methods*, 4th edn, Allyn and Bacon, Pearson Education, Boston, MA.

Liu, A.M.M. (1999) A research model of project complexity and goal commitment effects on project outcome, *Engineering Construction and Architectural Management*, **6**(2), 105–111.

Liu, A.M.M., Fellows, R. & Ng, J. (2004) Surveyors' perspectives on ethics in organizational culture, *Engineering, Construction and Architectural Management*, **11**(6), 438–449.

Oxford English Dictionary (2007) http://dictionary.oed.com.eproxy1.lib.hku.hk/cgi/entry/00315594?, accessed 10 July 2007.

Snell, R. (1995) Psychic prisoners? Managers facing ethical dilemmas: cases from Hong Kong. In: *Whose Business Values* (eds S. Stewart & G. Donleavy), pp. 143–163, Hong Kong University Press, Hong Kong.

Victor, B. & Cullen, J.B. (1988) The organizational bases of ethical work climate, *Administrative Science Quarterly*, **33**, 101–125.

Part 3

Reporting the Results

Chapter 9

Results, Inferences and Conclusions

> The objectives of this chapter are to:
>
> - discuss the requirements for valid results;
> - examine potential sources of *error* and *reliability*;
> - consider alternative ways of producing and presenting *results*;
> - examine the use of statistical *inferences*;
> - examine the requirements of *conclusions*.

Requirements for valid results

Once the research project has been structured, the theory and literature studied, the data collected and analysed, the next stages are to produce results and, by making inferences, examine and discuss the results of empirical work in the context of theory and literature, to draw conclusions and make recommendations; limitations of the study must be made explicit. The results relate to the analyses of data, whilst the conclusions use those results, together with the theory and literature, to determine what has been found out through the execution of the study. Particularly, conclusions must relate to any hypotheses proposed, the objectives set and, hence, to the overall aim of the research.

It is important to be sure of the validity and reliability of the work – the confidence which someone may have in the findings. One should judge how the findings may be used in other research and in application in practice. Part of such appreciation leads to recommendations for further research – this is identification of additional areas of study to extend and complement the work which has been carried out; it will inform the development of subsequent research projects. Thus, results are what emerge from analyses and, as such, require interpretation to give meaning in the context of what the research sought to discover. They must demonstrate what has been found out through the execution of the research.

For quantitative studies, statistical inference is employed to determine the applicability of the results to the issues under investigation and, thence, the drawing of conclusions. For those aspects, confidence (or significance) levels should be quantified and stated.

Example

In the production of concrete, slump tests are employed to test the conformity of the batch of concrete with the specified mix for workability and strength requirements. If the results lie within the limits prescribed, it will be inferred that the concrete will be suitable, and vice versa. Further, cube strength tests for concrete at 7 and/or 28 day strengths use comparisons of test results with design requirements over time to make inferences about the strength of the concrete tested.

Note, in both tests, the importance of following the correct, prescribed test procedures for inferences to be valid. Using a poker vibrator on site to vibrate a 150-mm concrete test cube may give results of very high strength, but those results are utterly useless for making inferences about the strength of the concrete placed on site because of the radically incorrect method of carrying out the test.

Conclusions take a 'broad perspective', looking at the research executed as a whole, but focusing particularly on the hypotheses, objectives and aim of the research, adopting an incremental approach to generalisations which may be made.

Almost inevitably, important issues will be identified during the course of the research – some will be incorporated into the study whilst others will remain outside its scope. The issues incorporated should be subject to conclusions, whilst those identified but not researched should be noted in the recommendations for further research.

Potential sources of error

Good research design attempts to control the sources of error (see Fig. 9.1). Total error – composed of random sampling error and non-sampling error – in quantitative measurement is the variation between the true mean value

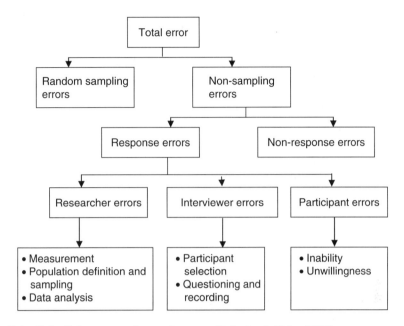

Fig. 9.1   Potential sources of error (*source*: Malhotra & Birks 1999).

in the population of the variable of interest and the observed mean value obtained in the research.

Random sampling error is the variation between the true mean value for the population and the true mean value for the original sample. *Non-sampling error* is attributed to sources other than sampling and can be random or non-random in nature, For example resulting from problem definition, approach, scales, questionnaire design, interviewing methods, data analysis and so on. Non-sampling error consists of non-response error and response error. *Response error* can be made by researchers, interviewers or participants, and it occurs when participants give inaccurate answers or their answers are mis-recorded or mis-analysed. *Non-response error* occurs when some of the participants included in the sample do not respond.

Hence, it is advisable to examine issues of validities (see previous chapters) and reliability.

Reliability

Researchers can never know for certain precisely how much measurement error is contained in a respondent's score or what the true score really is. If the researcher combines the scores of many respondents and calculates the variance, the total variance of the set of scores is composed of the same two components:

Total variance in a set of scores = variance due to true scores (systematic variance) + variance due to measurement error (error variance).

There are three common ways to estimate the reliability of the measures: test–re-test reliability, inter-item reliability and inter-rater reliability. If two measurements of something yield very different scores, the measures must contain a high degree of measurement error. Test–re-test reliability is determined by measuring participants on two occasions, then the two sets of scores are correlated to see how closely related the second set of scores is to the first. Inter-rater reliability involves the consistency among two or more researchers who observe and record participants' responses. Inter-item reliability is relevant for measures which consist of more than one item and assesses the degree of consistency among the items on a scale (internal consistency reliability). Internal consistency reliability is used to assess the reliability of a summated scale where several items are summed to form a total score. In a scale of this type, each item measures some aspect of the construct measured by the entire scale, and the items should be consistent in what they indicate about the construct. This measure of the reliability is the coefficient alpha which focuses on the internal consistency of the set of items forming the scale, for example Cronbach's alpha.

> *Example*
> Cronbach's alpha is a measure of the internal reliability of the items on a test or questionnaire and ranges from 0 to 1.0. It reflects the extent to which the items are measuring the same thing, for example if all participants who said 'no' to question 1 also 'no' to questions 2 and 3, the alpha for these three would be 1.0.

Results

Producing the results

Results record the outcomes from tests. The selection of appropriate tests to analyse data is very important. In some cases, a variety of analyses may be employed, both statistical and/or non-statistical. Given sufficient time and other resources, it is useful to employ 'triangulation', using a variety of analyses of the data so that the results which are produced can be considered both from the viewpoints of the individual analyses and from the perspective of the combination of the analyses. In particular, attention can focus on the analyses which produce results which agree with each other broadly, if not exactly, and any which produce conflicting results. Not only are the results dependent on the tests which have been carried out, but are dependent upon the data which have been collected and the recording of those data and their coding, if applicable.

Naturally, for some 'practical' purposes, small errors in coding data may not be significant but, for any type of research, errors are likely to be material. Thus, elimination of errors requires both careful design of the coding system to ensure clarity and ease of allocation, plus checking of the allocations. Apart from mistakes in allocations, ambiguity is the main source of errors. Such misallocations will, at least, distort the results.

Triangulation may be carried out by collecting several sets of data. Often, the data will be collected from different samples, although sub-sets of primary samples may be used – such as where interviews are conducted with a sub-set of respondents to a questionnaire survey sample. In such circumstances, it is important that an appropriate, unbiased method is used to select the sub-set. The sub-set should be selected either by random sampling or by following the sampling frame. Naturally, this will be limited by those respondents who do not agree to participate in interviews and so, the issue of non-responses must be addressed.

Introductory results

Commonly, research projects yield a considerable number of individual results from the analyses of the data. Some results will be simple – in most cases, introductory results will be needed to provide a suitable structure for the data employed for the analyses.

Assuming that the population size and structure, types and number of cases or sampling frame, methods and samples sought have been noted, introductory results to describe the data analysed for each set and sub-set should include:

- useable response numbers and rates (expressed as percentages);
- descriptive statistics *relevant to the investigation* – for example, sizes of organisations, mean turnover, standard deviation of turnover;
- description of individuals responding – such as job title (to indicate nature of views provided and credibility and authority of those views).

Such results would complement the discussion of how the data were sought by describing the actual sample analysed compared with the sample desired. For laboratory experiments, the experimental technique, equipment

and so forth will be described and discussed whilst, for modelling and simulations, the sources and nature of the data and information and the methods
for building the model or conducting the simulation will be noted.

Minor differences between samples desired and those obtained are unlikely
to be of great importance. However, if samples sought are small and, given
that tests on smaller samples are usually less reliable than tests on larger
samples, small numbers in the samples will mean quite large percentage
differences – it is the proportionality of differences which tends to be important. Thus, small numerical differences may cause samples available for
testing to be so small that any results may be (almost) meaningless. In situations of quantitative analyses, particular tests are available for use with
small samples.

The second aspect to examine is the pattern of the differences. If the differences are random, or follow the sampling frame proportionately, they are of
less significance than if they follow a particular, different pattern. Where this
occurs, and the differences show bias in themselves, they introduce bias into
the sample analysed and, hence, the results obtained. So, differences between
samples desired and those obtained may affect the validity of the results, at
least in terms of confidence, due to size of data sets useable and considerations
of bias.

It is useful to determine why non-responses have occurred and, if possible,
to investigate whether groups of non-respondents differ in their views and
so forth from the respondents and from other groups of non-respondents. For
low response rates and small response numbers, the views of non-respondents
will be dominant but, without appropriate investigation, unknown and so, the
validity of the research findings is questionable. In this situation, it may be
advisable to constrain the sample and its size and to review the data collection
method such that response rates from the revised sample will be sufficient to
ensure only small possible non-response bias. Apart from the issue of non-
response bias, significantly lower response rates than those anticipated may
mean that statistical testing may require modification from tests valid on
large samples to those applicable to small samples. The boundary between
'large number' and 'small number' statistics is at $n = 32$, although the size
adopted in practice often is $n = 30$ (see Levin & Rubin 1991), where n is the
sample size.

Substantive results

Having provided a description of how to subject the data to analyses to
yield the results, the substantive results themselves may be considered. For
experiments, modelling and simulation, sources of possible errors should

be noted and, as far as possible, quantified, to yield an overall measure of 'experimental error' (i.e. a variability/reliability measure). Although certain results will be sufficient in themselves to provide the required information, others will need aggregation, companions and rearrangement to enable researchers to maximise understanding of what the results mean. Different forms of presentations will be useful also – notably charts and graphs; tables may convey precision but are much more difficult to interpret.

Especially for applied research projects, in which objectives will have been set at the outset, and where the research is targeted towards the provision of answers to particular questions or issues, the results, and the research leading to them, can be categorised as follows (following Ritchie & Spencer 1994):

- Contextual: identifying the form and nature of what exists
- Diagnostic: examining the reasons for, or causes of, what exists
- Evaluative: appraising the effectiveness of what exists
- Strategic: identifying new theories, policies, plans or actions.

Clearly, there should be a fairly 'seamless' transition from results of analyses to the conclusions drawn via discussion of the results in the context of the theory and literature. Often that transition leads to lack of appreciation of the differences between results and conclusions. The differences, however, are quite marked – results are produced from analyses of data, whilst the conclusions consider the results in the contexts of the topic and the whole of the study in order to determine what exists, why and how, and the considered consequences of issues examined in the research.

For projects which obtain qualitative data, the analysis will initially have involved the researcher's getting 'immersed' in the data to gain maximum familiarisation; the first of the five analytic stages noted by Ritchie and Spencer (1994). The second stage is identification of a thematic framework followed by indexing, charting and, finally, mapping and interpretation. At this last stage, the key objectives of the analysis are addressed. In such studies, the 'development' of results via analyses is more appropriate a view than the 'production' of results from analysis, which is appropriate in quantitative studies.

The stages of charting and mapping, and interpretation involve the production of results. The charts include headings and sub-headings and are laid out 'thematically' – by themes across responses, or by case – for respondents across the themes. Mapping uses the charts to extract key characteristics from the data. While the charts and analytic processes leading to their production present the raw results, mapping enables the researcher to juxtapose alternative views, to set out pros and cons such that the charts present evidence which can be structured through mapping.

Inferences

Inference is the process by which the meanings and implications of the results are determined in order that conclusions may be drawn. To assist the process of inference, the form and levels of aggregation of the results must be considered. Most usefully, the transition from results to conclusions of a research project is effected in the report by a section or chapter involving discussion of the results. The results will be discussed in the contexts of each other and of the theory and literature. Throughout, it is appropriate to ensure that the discussion is relevant to the pursuit of the aim and objectives of the work although it is usual for other issues to emerge.

Whetten (1989) notes that '... the essential ingredients of a simple theory ... [are] ... description and explanation' and 'data, whether qualitative or quantitative, characterize; theory supplies the explanation for the characteristics'.

In order to gain full understanding of the results and appreciation of their implications, it is helpful to present the results in a variety of ways, including tables and diagrams. The diagrams provide immediate indications of generalities, trends and so forth, whilst the precision of detail contained in tables is important to facilitate particular considerations and subtleties.

In considering applied policy research using, particularly, qualitative data, Ritchie and Spencer (1994) note that, in seeking associations, '... the analyst may become aware of a patterning of responses ... it may appear that people with certain characteristics or experiences also hold particular views or behave in particular ways ... the analyst will systematically check for associations between attitudes, behaviours, motivations etc. ...'. Analogous considerations apply to quantitative results.

By inference and discussion, researchers seek to gain insights into how the individual parts of the study relate to each other in terms of the issues which the individual parts identify. In the absence of theory, results are isolated; their consideration in the light of theory and previous research findings facilitates the advance of knowledge and a perception of how the topic and its practice is developing.

Causal relationships

Results must be interpreted in the context of each other, of the theory and of results of previous research. Such interpretation is intended to reveal relationships between the results in terms of extent and, it is hoped, direction, as well as helping to gain insights into causation.

Generally a good theory describes and explains relationships between constructs and variables within specified boundaries and, further, facilitates prediction. Thus, predictions based on theory are grounded in the propositions of the theory and the hypotheses which have been deduced and tested and so, the predictions must be qualified by reference to the constraints – usually regarding circumstances, time and/or number of instances. Predictions based (solely) on probability assume that, given enough time and/or instances, all outcomes may be predicted probabilistically. Thus, the power and importance of theory lies in its ability to explain (causality).

A difficulty in investigating causal relationships between pairs of variables is the potential causal influence of other variables. Moser and Kalton (1971) consider how the effects of such further variables can be dealt with in the evaluation using multi-variate analysis techniques. A more fundamental notion is to employ theory to examine the logic of causation.

Example

There may be a correlation between phases of the moon and, lagged by the appropriate time interval, births of babies; such correlation does *not* demonstrate causality and what knowledge is available indicates very strongly that it is something other than the phases of the moon which causes births of babies.

In some instances, although theory may suggest direction and causation of relationships between variables, it may not be conclusive. This is quite common in the social sciences; for example, it is not clear whether demand leads supply or vice versa and, perhaps, both relationships operate in the dynamics of a market economy. Further, a variety of theories may be appropriate – such as those relating to inflation, for Keynesian, Monetarist theories and so on.

Hence, the discussion of results in the context of theory and literature allows alternative considerations and views to be examined. Results should indicate strengths of relationships – often epitomised by the confidence level, determined by the degree of statistical significance. They should also provide the numerical statistical measure of the relationship, taking sample size into account. Thus, inference requires an holistic view of the research project to be adopted.

In studies which incorporate both qualitative and quantitative data and analyses, it is essential to bear in mind that the data are founded on different logical principles and, consequently, so are the results. Therefore, it is important to be aware of both the principles and the sets of assumptions, which should be explicit, in order to assist the researcher to move between the two sets of results with ease.

Interpretation

Inferences and discussions enable the researcher to present the issues arising out of the research from two perspectives separated in time – that prior to the execution of the empirical work and that following its execution and production of results. Comparison of the two perspectives is important to demonstrate how knowledge has changed due to the study – to reinforce or to question the previously 'perceived wisdom'.

In interpreting results, associations and causalities between variables are investigated. Usually, variables are considered in pairs; independent and dependent variables. In doing so, 'third' variables must be eliminated by the sampling approach adopted or by adjustment in the analysis, as demonstrated in the following example.

Example

A random sample of 2000 people who smoke is selected to investigate whether a short TV campaign will induce them to give up smoking. Six weeks after the end of the campaign, they are asked whether they have given up smoking:

	Viewed (V)	Not viewed (N)	Total
Still smoking (S)	500	300	800
Given up (G)	1000	200	1200
	1500	500	2000

67% of those who viewed the campaign had given up smoking whilst only 40% of those who did not view the campaign had given up. The results suggest that the campaign was successful but other factors (variables) could have been influential – consider 'social class' and 'age group'.

	(V)	(N)	Middle class (M) (V)	Middle class (M) (N)	Working class (M) (V)	Working class (M) (N)
(S)	500	300	125	75	375	225
(G)	1000	200	250	50	750	150
	1500	500	375	125	1125	375

Analysis by 'elaboration' of the 'social classes' variable yields the same percentage of viewers and non-viewers who gave up smoking in both of the social classes as in the total sample. Hence, 'social class' is not a significant variable.

continued

Example   Continued

	Younger (Y)			Older (O)		
	(V)	(N)	(V)	(N)	(V)	(N)
(S)	500	300	275	225	225	75
(G)	1000	200	475	25	525	175
	1500	500	750	250	750	250

In the older age group, 70% of those who viewed and 70% who did not view gave up smoking, whilst in the younger group 63% of those who viewed the campaign gave up smoking whilst only 10% of the non-viewers in that age group gave up.

Clearly, age group is significant. 70% of the category 'older people' gave up smoking irrespective of whether they viewed the TV campaign, whilst the campaign was significant for the younger people in inducing them to give up smoking. Thus, age is a moderator variable but social class is not.

Associations *within* groups are *partial* associations and *between* groups are *marginal* associations. For the social class analysis, marginal associations of social class with viewing the TV campaign and with giving up smoking do not explain any part of the overall association; the association is explained by the partial association completely. This is termed *elaboration by partials*. For the age group analysis, marginal and partial associations are apparent and so, both types of elaboration are relevant.

In interpreting results (and using data), Tversky and Kahneman (1974) noted that the following human behavioural aspects are important.

Representiveness considers relationships in the form that α is probably related to A, if A is a category (population; type of process) and α is some particular instance. Thus, if A typically comprises α's (or produces α's from the process), then the usual heuristic-based judgement is likely to be biased towards α being related to A. Further information about the potential relationship tends to be ignored.

Hence, people tend to ignore (be insensitive to) prior probability of outcome (or base-rate frequency) and so, classify people, events etc. on the basis of conformity with stereotypes rather than prior probability (e.g. the percentage of architects in the population).

Further, people are insensitive to sample size – small samples are much easier to collect but are more likely to be distorted from representing the population than are large samples (e.g. sensitivity to extreme values).

Probability is often misconceived, such as where sequences appear to be systematic rather than random; for example, in tossing a fair coin, the nth toss has $p = 0.5$ (h) and $p = 0.5$ (t) irrespective of what occurred in the $n - 1$

tosses. Although many believe that randomness acts to 'correct' deviations (in this example, a sequence of tosses coming up heads), in fact, it merely dilutes (reduces) the effect on the overall situation (such as the distribution of the results of a sample of tosses of the coin) as the process (the tossing of the coin) continues.

A particular research issue concerning sampling (size of sample) in survey designs is a belief in the 'law of small numbers' which, erroneously, asserts that even small samples provide good representations of populations. Proper consideration of calculated confidence levels of statistical measures, which, usually, depend upon the size of the sample (as in, e.g. coefficients of correlation), should help to dispel that illusion.

Insensitivity to predictability suggests that people are likely to predict an outcome based on the favourableness of the descriptive information given and will not tend to take the reliability of evidence or the expected accuracy of the description into account.

The illusion of validity follows on from the prediction using representativeness. If the degree of representativeness is high, people have high levels of confidence in their predictions, irrespective of factors that limit prediction accuracy. Therefore, people tend to fit their prediction to the input (base) information.

Regression effects may not be well appreciated in that successive tests on a sample yield 'regression towards the mean' of the population. This is an analogous manifestation to larger samples being more representative than small samples.

Availability concerns subjective assessment of the probability of an event through ease of recall. Instances of frequent (large) classes are probably recalled better and faster than instances of less frequent classes. Hence, reliance on availability leads to predictable biases.

The biases occur through retrievability of instances, notably familiarity. The bias is enhanced by 'salience' – observing a dramatic event has more impact than reading about the same event. Further, the easier the search pattern makes the searching for results, the greater the impact (+ve) on the outcome. Also, the more easily imagined a future event is, the greater the positive bias in assessing that event's likelihood (probability). Further, it appears that if some aspect of 'common sense' suggests that a correlation is likely, then such an illusion will bias results.

Adjustment and anchoring are important because a very common aspect of forecasting is to select some (suitable) starting point and then adjust that value to produce the result (prediction). Commonly, adjustments are not adequate – results are biased towards the starting value. Hence, selection of the starting value takes on much more importance. The initial problem is that of insufficient adjustment. (As in the selection of items from a priced BQ for a project to use as the basis for calculations of pro-rata rates in valuing variations.)

People tend to overestimate the probability of conjunctive events – a sequence of events (the product of the individual component event probabilities) – and underestimate the probability of disjunctive events – probabilities of component events are summed. Thus, a conjunctive event has dependent components and a disjunctive event has independent components.

Many complex events, such as completion of a construction project, comprise conjunctive component events which have to be completed sequentially. Even if successful completion of individual component events is highly probable (e.g. duration prediction), the probability of successful completion of the total project may be low. There is a distinct tendency to be overoptimistic in forecasting the (probability of the) overall event – particularly if non-probabilistic methods are used. (Consider the effects of 'merge events' in calculations of the probabilities of overall durations predicted for a project when using PERT programming.)

In any complex system, such as an excavator, the system fails if any individual but vital component (e.g. a hydraulic hose) fails. Even with very low probabilities of failure of each of the vital components, system failure probability may be high.

If people are requested to judge the variability of their forecast, it is common for the boundaries to be expressed to be more narrow – closer to the forecast – than is justified from objective assessment. Such constriction of the subjective probability distribution of the forecasts is another aspect of anchoring.

Given the above, Tversky and Kahneman (1974) concluded that 'These heuristics are highly economical and usually effective, but they lead to systematic and predictable errors!'

Tversky and Kahneman (1974) suggest that forecasts which incorporate much human judgement frequently operate through the use of '... a number of heuristic principles that reduce the complex tasks of assessing probabilities and predicting values to simpler judgemental operations ... [and so,] ... sometimes lead to severe and systematic errors', as found by Reugg and Marshall (1990) in the contexts of construction price forecasting.

The potential sources of judgemental bias in building price forecasts (due to problems of representativeness, availability, and anchoring and adjustment) are examined by Gunner and Skitmore (1999). Briefly, representativeness problems are likely to encourage single figure forecasts as mean values (etc.) are afforded too much weight, similarity is overemphasised and random events are regarded as systematic. Availability leads to emphasis on large factors and frequent occurrences. Anchoring leads to weighting of forecasts towards the initial data (e.g. price per m^2 from a database). Anchoring bias is extended by adjustment which results in underestimation of the degree of difference from the original data.

Kahneman *et al.* (1982) conclude that 'people tend to be both inconsistent and biased in their interpretation of information'. Hence, Chau *et al.* (1998) assert that 'The important issue is that at each stage the [methodological] approach is rigorous in that the issues are defined clearly and the logic of the argument is made explicit along with any assumptions implicit in the approach adopted.'

Conclusions

Conclusions are the major output of the research – what has been found out about the topic under study through the execution of the research. As such, conclusions present the major items and themes which have been raised and investigated. Conclusions complete the 'story' of the research from what it was desired to *find out* to what has been *found out*. It is neither appropriate nor useful for (apparent) conclusions merely to summarise what was done in the study nor to just repeat the statements of results. Conclusions must be concise statements of the contributions to knowledge which the research makes.

How to write conclusions

Generally, a research project will yield between six and twelve main points of conclusion – those facets about which the study has helped to develop knowledge, what those developments are and what they mean for practice and further research work. Ideally, each conclusion stands alone, usually as a paragraph, but with sure and clear foundations in the research which has been executed. Only if, coupled with the theory and literature, the empirical research carried out leads to the conclusion should it be stated as a conclusion; mere opinions, whims and conjectures of what could or might be are not appropriate. As grounded theory leads to the development of theories and so on by scrutiny of data collected and identification of patterns and relationships, so the conclusions of any research project must truly emerge from the research actually done; no new material should be introduced in the conclusions.

A common good practice is to produce one stand-alone, often numbered, paragraph for each point of conclusion. For studies which set out explicit aim, objectives and hypotheses, it is important that the conclusions relate back to them. Hypotheses tested through the research must be subject to conclusions. These should state whether the research leads to support or rejection of the hypotheses, with what level of confidence, why, and with what likely consequences, given the other results and conclusions. The overall aim was used to determine objectives and, thence, hypotheses through analysis. So, synthesis of the conclusions regarding the hypotheses leads to conclusions about the objectives which, in turn, relate back to the aim. This approach facilitates consideration of the degree to which the objectives and aim have been realised, what further research should be done, how future studies of the topic may be executed and, by examining the validity of the study and its findings, what may be recommended for implementation and for further research.

It should be possible to reference each statement in each conclusion to a section of the research. Although such tracing should not appear in the report, it may be useful to carry out the exercise as a mechanism for arriving at the conclusions and ensuring their substantiation by the research done. In quantitative and applied studies, significant assistance in determining the topics to be addressed in the conclusions is given by the presence of a detailed aim, objectives and so on virtually from the outset. In such circumstances, those specified themes may be tracked through the work so that the major issues are addressed progressively in the theory, literature and empirical sections – this is of much assistance in developing and drawing conclusions.

Occasionally, it is useful to draw 'sub-conclusions' on completion of each section of the research project. These conclusions help to isolate what has been substantiated, what has been refuted and what remains as unresolved issues. Especially when new topics and qualitative work of investigation are being pursued, such a progressive approach can be very helpful in developing the issues to be studied as the research progresses.

In qualitative studies, the drawing of conclusions tends to be far less 'directed' due to the usual absence of detailed objectives and hypotheses; indeed, it is common for qualitative studies to set out to produce theories which are intended to inform objectives and so on for subsequent research. Hence, conclusions from qualitative studies may be in the form of identification of relevant variables, patterns and relationships between those variables. These may be either discovered by the research or by 'logical' but hypothetical issues to be the subject of future investigations. As qualitative studies tend to be concerned with research into new topics, analogies with relationships in other topics and subjects may not apply. What may be logical and rational in one discipline may not apply elsewhere, if only because different sets of assumptions are employed. Further, in moving between different groups of people, behavioural factors may alter considerably.

> *Example*
> Logical, rational behaviour in one industry may not be appropriate in others. For example, newspapers and property investment; newspapers must take a short-term view in most instances whilst property investment is a long-term activity.

It is essential to be aware of the variables and the assumptions as well as the principles, samples and results of analyses in drawing conclusions.

Further research

It is unlikely that any study can be comprehensive. Most hope to meet the majority, if not all, of the objectives, but that achievement is somewhat less than realising the aim of the research. This leads to three further facets of the research. In particular, examining the conclusions and comparing them with the aim and objectives will suggest:

- recommendations for implementation;
- limitations of the study;
- recommendations for further research.

Evaluation of what is concluded with high levels of confidence from the study leads to any recommendations for implementation – particularly appropriate for applied research. Note what should be implemented, how and why.

Limitations explain why the scope of the study, results and so on was constrained. Certain limitations are likely to have emerged during the course of the research – not everyone who agrees to respond to a survey and so forth will always do so; other circumstances change too, favourably as well as unfavourably. The limitations, especially those occurring during the research and outside the control of the researchers, are important to note with their consequences. Research is dynamic and occurs in a dynamic environment hence, change is inherent.

Recommendations for further research should suggest topics for study, appropriate methodologies, given the knowledge gained from the research just executed, and the reasons why such further studies would be useful. Mere recommendation of replicating the study with a larger sample which might reinforce the results is hardly a useful recommendation – it's obvious.

The transition from results to conclusions and recommendations through inferences requires *insight* – the conclusions should express and explain those insights such that the conclusions and recommendations will be *informative*.

Summary

The chapter has considered various approaches to the production and presentation of the results obtained from analysing the data collected. Consideration of the results in the context of the theory and literature enables inferences to be drawn and the results to be discussed in context. It is important that conclusions are drawn from the research carried out and are not mere ideas or whims. The conclusions state what has been found out during a particular research project and should relate to the aim, objectives and hypotheses, if any. Notes of the limitations of the study and recommendations for further research and for appropriate implementations should be made.

The research report presents a story in three parts:

- what the study seeks to find out;
- how the study tries to find it out;
- what the study has found out and what future studies might attempt to find out.

References

Chau, K.W., Raftery, J. & Walker, A. (1998) The baby and the bathwater: research methods in construction management, *Construction Management and Economics*, **16**(1), 99–104.

Gunner, J. & Skitmore, R.M. (1999) Pre-bid price forecasting accuracy: price intensity theory, *Engineering Construction and Architectural Management*, **6**(3), 267–275.

Kahneman, D., Slovic, P. & Tversky, A. (eds) (1982) *Judgement Under Uncertainty: Heuristics and Biases*, Cambridge University Press, Cambridge.

Levin, R.I. & Rubin, D.S. (1991) *Statistics for Management*, Prentice-Hall, Englewood Cliffs, NJ.

Moser, C.A. & Kalton, G. (1971) *Survey Methods in Social Investigation*, Dartmouth, Aldershot.

Reugg, R.T. & Marshall, H.E. (1990) *Building Economics: Theory and Practice*, Van Nostrand Reinhold, New York.

Ritchie, J. & Spencer, L. (1994) Qualitative data analysis for applied policy research. In: *Analysing Qualitative Data* (eds A. Bryman & R.G. Burgess), pp. 173–194, Routledge, London.

Tversky, A. & Kahneman, D. (1974) Judgement under uncertainty: heuristics and bias. *Science*, **185**, 1124–1131.

Whetten, D.A. (1989) What constitutes a theoretical contribution?, *Academy of Management Review*, **14**(4), 490–495.

Chapter 10

Reports and Presentations

The objectives of this chapter are to:

- outline the essentials for good *report production*;
- discuss the importance of effective and efficient *communication*;
- examine the *contents of the report*;
- note the essentials of *oral presentation*.

Report production

Having completed the research, it is essential to produce a report of what has been done and what has been discovered, so attention must be given to its content and form. In many instances, other forms of presentation will be required as well. It is usual for the style, layout and, sometimes, length of a research report to be specified by the course documents, the university or the organisation commissioning the research.

The report of a research project is the primary source from which the research will inform 'the world at large' of what has been done and what has been discovered. It is a primary communication document and, hence, is of paramount importance. If the communication does not work well, the research may never realise its potential so it is essential to reflect on how to achieve good communications at the outset of producing the report. Even academic reports can be lively and stimulating if presented well, despite the possible constraints of required format, rigour and academic style. Graphics and colour can enhance the message. It is important that the report is clear and concise; it must denote what has been done, why and how, to a level of detail to facilitate both extension of the work and its replication; an approach employed widely in experimental work.

Naturally, everyone has their own approach to writing a report. However, the report of a research project is likely to have the material assembled over a significant period of time, from several days to several years. The study may have been undertaken by one person alone or, more usually, by two people (such as researcher and supervisor) or, as is becoming increasingly popular, by a group of people, possibly drawn from a variety of disciplines and, hence, research traditions. As considered in reviewing the theory and literature, it is essential that adequate records are maintained. Some people elect to write draft sections of the report as the work proceeds which may assist distillation of the main ideas and issues; others prefer to keep notes and to write the report as an entity once the research work has been finished, the latter approach may aid continuity of themes. What is clear is that there is no universally best way to produce a report; although certain principles may be applied, the selection of the process is at the choice of the individual(s) and so, subject to that person's preferences.

Assistance and guidance of the supervisor (or team leader), if there is one, can be very significant. The approach to producing the report should be discussed and agreed with the supervisor – preferences and alternatives should be considered with the goal of securing the best report within the constraints – notably time. For researchers who are not good at producing and adhering to deadlines, it is likely to be a good idea to set and agree a schedule for the production of the report and for the supervisor to 'police' compliance with the schedule. Remember to incorporate some contingency time for revisions and so on. It is a good idea to study a few research reports which in topic, level or/and purpose are similar to the one to be produced. It may be useful to do this early in the research period – at proposal formulation – to gain awareness of the academic 'target', as well as at the report production stage. An initial draft outlining the contents of sections or chapters should be submitted to the supervisor, and also to colleagues if they are willing, to ensure that what is produced is what is required. When drafting the final report, it is a good idea to submit the first portion of the report once drafted to the supervisor and not to write more until agreement has been reached that the portion submitted is satisfactory in terms of writing and presentational style, and scope and depth of content; usually, that portion will be (part of) the review of theory and literature.

Communication

Any communication exercise involves two parties or groups – a sender and a receiver, as shown in Fig. 10.1.

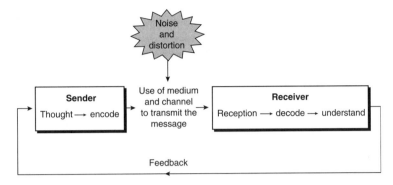

Fig. 10.1   Communication system.

The aim is to transfer the thought of the sender as accurately and, usually, as quickly as possible to the mind of the receiver. Various languages, at least one, will be employed in the process, so common understanding of the language is essential. If translation is required, say from Cantonese to English, and neither the sender nor the receiver speaks both languages, an intermediary link in the communication chain must be introduced; the translator, through the translation process, will 'filter' the message whilst acting as both receiver and sender of each element of the total message being transmitted.

Clearly, noise and distortion are likely to occur to some degree in any communication; noise is interference from the environment, such as a machine operating close to where a conversation is taking place, distortion is interference in the communication system, for instance an echo on a phone line. Bias can occur in the way messages are phrased; contrast, 'Do you want to eat?' with 'Do you really want to eat?'

Feedback, an essential element of any system, to facilitate evaluation of efficiency and effectiveness, is, itself, a communication chain and so, is subject to the potential problems of any communication system. Communication chains should be kept as short as possible, multiple forms of communication which reinforce each other in transmitting the message are useful. The most important consideration in communicating any message is the ability of the receiver to receive, decode and understand the message accurately and readily as the sender intends. If that understanding cannot be achieved, the communication has failed to a degree; the greater the difference between the intended understanding by the sender and the actual understanding by the receiver, the greater is the failure. Feedback may occur through interpretation of the actions taken by the receiver once the message has been transmitted.

The report of the research project is the primary communication to assessors, commissioners of the work, colleagues and others, of what was done, how, when, why and what discoveries were made. Usually, the basic form of the communication is prescribed, but other communications will occur throughout the research activities; notably with the supervisor during the production of the report.

Given the statement of requirements and limitation parameters for the report, it is vital to consider its contents and how they may be presented best. It is important to write clearly, concisely, avoiding jargon and pomposity. Some writers seem to believe that it is silly to use one word when a hundred will do. Oscar Wilde apologised for writing a long letter as he did not have time to write a short one . . . Leave time for reviewing the report and editing it.

Contents of the report

A research report should include:

- Title page
- Abstract
- Acknowledgements
- Contents list
- List of figures
- List of tables
- List of abbreviations
- Glossary of terms
- Text of the report
- References
- Bibliography
- Appendices

How to begin

The title page may be required to follow a prescribed format. It should state the title of the research project together with the author(s), the purpose of the research, such as the partial fulfilment of the requirements for the award of a degree, and the date of completion and submission of the report.

The abstract is a very concise summary of the research report, usually in 250–500 words. It should outline the topic, briefly state the aim, main objectives and hypothesis, summarise the methods of data collection and analyses and state the main findings and conclusions. It is best if the abstract is the final item of the report to be written to ensure that it does abstract/summarise the content of the full report very concisely.

The acknowledgements page is the opportunity to include brief, formal statements of thanks to people who have helped in the execution of the research – mentors, supervisors, providers of data, fellow researchers, secretaries, editors and so forth and, if applicable organisations/agencies which have provided financial or other forms of support. Whilst it is polite to thank everyone who has assisted with the work, it is best not to be overzealous in scope or expression and it is *most important* to ensure that confidences and anonymity assurances are not betrayed unwittingly.

The contents list, supplemented by lists of the figures, diagrams and tables, should note the chapters, main sections and sub-sections of the report, plus any appendices and so on with page numbers. The list provides an 'at a glance' overview of the report for readers; it is an invaluable guide for anyone who wishes to obtain particular facets of information without having to search the entire text.

The list of abbreviations, is a useful guide to ensure that both 'standard' and 'particular' abbreviations are understood by all readers, and that the author and readers have a common understanding. All abbreviations used in the report should be listed alphabetically with the full items noted against each abbreviation. The glossary of terms is included to ensure common understanding between author and readers. Terms which have particular meanings in the subject topics of the research and in industrial or practice context must be listed alphabetically with their definitions.

Text of the report

The text of the report constitutes the bulk of the document. Although chapter headings, sequences and breakdowns of contents will vary, notably between quantitative and qualitative research, a significant amount of contents are 'standard'. The text will begin with an introductory chapter which will discuss the topic of research, the rationale for the work – based around the research questions or problem(s) which the research addresses, the aim and objectives. Commonly for quantitative research, the introduction will state the main hypothesis and outline the research methodology and methods. Often, it is useful to include a diagram of variables and hypothesised relationships between them (a nomological network). It may be helpful to end the introductory chapter with a brief summary of the contents of the other chapters in the report and appendices, but care should be taken that the brief summaries are not just repetitions of the contents list.

Given a thorough proposal and having executed the work, planning the content in the text of the report is likely to be quite straightforward. The first and last chapters are 'Introduction' and 'Conclusions and Recommendations' *but*, in order of writing, these are the final and penultimate chapters produced, respectively. This is because the conclusions and recommendations relate to the entire work, which is outlined briefly in the introduction.

The remaining content in the text of the report is likely to be written in the chronological sequence in which the research was executed – Theory, Literature, Methodology and Methods (unless sufficient detail has been incorporated in the Introduction), Data Collection, Data Analysis, Production of Results, Discussion of Results. How the activities are organised into chapters is one of appropriateness.

The suggested order of writing the text of the report is:

- Theory
- Literature
- Methodology and methods
- Data sourcing and data collection
- Presentation of results
  - Data analysis
  - Production of results
- Discussion of results
- Conclusions, limitations of the study and recommendations
- Introduction
- Abstract

Theory and literature

For much research, theory is the 'bedrock' component. Although originally derived through research, theory constitutes the body of knowledge which has secured general acceptance through rigorous testing which has corroborated the contents (variables, constructs, relationships) and so, much theory has achieved the status of 'laws' of the subject, thus, it is used to inform and guide further work. All potentially, as well as actually, relevant theory should be considered and, therefore, critically reviewed and summarised in the report. The next element is the review of literature – the summary, juxtaposition and evaluation of the samples, methods, results and conclusions of research executed already by others but which has not been sufficiently corroborated to constitute part of theory. There must be reviews of theories pertaining to the subjects of study and a critical review of literature yielding a reasoned evaluation of what is known relating to the research topic. The theories are basic laws, principles and so on which underpin the subjects of the research, whilst the literature is reports of researchers into developments and/or applications of theories. Such reviews may constitute one or several chapters depending on the purpose and 'level' of the research, the topic and the scope of underpinning subject material.

Theory and literature may be combined, but they are summaries of the work of others and so, must be *fully and properly referenced*. As this is a review of the (past) work of others, they must be cited in the text wherever their output is quoted or paraphrased and so, the section should be replete with citations. Intentionally not giving citations – and so, passing off the work of others as that of one's – own is fraudulent, intellectual theft known as plagiarism, which, properly, has very serious consequences!

The job of the researcher in producing the report is to present the salient points succinctly, demonstrating strengths of alternative views and issues, summarising what is accepted and what is subject to debate. Only from a comprehensive, critical review should the researcher express an opinion on the weight of evidence leading to a particular line of work, view or approach being adopted. The researcher fulfils the role of an informed, thorough, objective reporter; certainly not a 'tabloid journalist'. The task in this section is to report the work of others, not to do 'a sales job'; any expressions of the researcher's view *must* be supported by the weight of evidence presented in the report. The essence is to produce a critical review – the criticism must be both objective and informed. Each statement or set of statements should be referenced – opportunities for the researchers to express their own opinions will arrive soon enough.

In some studies, objectives for any empirical work, and hypothesis to be tested empirically, may not be developed until the reviews of theories and literature have been completed. In such cases, a brief chapter considering their development, with rationale, should follow the theory and literature review chapters. Particularly where the researcher is striving to formulate objectives and hypotheses, perhaps because of novelty of the topic of the study, the discipline required to produce the draft write-up of theory and literature can be invaluable in assisting objectivity and analysis of the 'state of the art' in the topic. Having carried out that exercise, it should be much easier to detect the main issues and so, to decide what aspects to research and how to proceed. Irrespective of the research approach adopted, the production of a list of issues, either within or separate from the report, is helpful to demonstrate how the particular research project fits into the debate. It also helps to advance knowledge of the subject and guides the recommendations of what further research should be done.

Almost inevitably, a researcher gets very close to, and possessive of, the research. Review by colleagues, and, especially, by the supervisor, is very helpful as the greater distance and consequent added objectivity ensures that what is written states what the researcher intended, and assists in the compilation of a full list of important issues.

Reporting on methodology and methods

In reports of larger research projects, it is likely to be appropriate to include a separate chapter examining methodology issues and providing the rationale for the adoption of the methods used. Such a chapter may follow the introduction or, usually more appropriate, be located between the chapters of theory and literature review and those which explain empirical work.

Methodology may be regarded as analogous to strategy in that it concerns the paradigm adopted (or followed) in the research; it relates to the research as a whole and so, includes the methods adopted for the various research activities. Recall the aspects of DATA discussed in Chapter 2. The selection of the most suitable methodology must be justified expressly in the report, unless the research concerns one of the few topics for which only one approach is suitable. The ontological and epistemological perspectives of the researcher underpin the selection of the research paradigm and, hence, the methodology.

Methods are operational approaches – the techniques which are selected and used to source data, to collect data and to analyse the data collected. Again, it is essential to express the rationale underpinning the selections made, usually comprising comparisons of both the academic and practical considerations – evaluation of the *pros* and *cons* of each method appropriate to the research. It is important that sufficient detail is included so that the choices are justified clearly and objectively. For analyses, it is important to demonstrate why particular tests were carried out – usually relating to the nature of the data, with references to the scales of measurement adopted, and to the research questions/ problems being addressed (as manifested in the objectives and any hypotheses).

Reporting on data sourcing and data collection

The next sections of the report to be written describe and discuss (justify) the sourcing of data; what data were collected, why, how and what the data collection involved. A discussion of the data obtained compared with the data desired is useful – if differences are minor, a note of those differences may suffice, but where differences are larger, consideration of possible and known reasons should be provided. Such a discussion should be carried forward to the consideration of results, and for examining the validity of the findings and limitations of the research. Response rates markedly different from the norms should be discussed; techniques which have produced really high response rates could be very useful to researchers in the future. Discussion of piloting, and evolution of methods due to data 'issues' should be highlighted in the report.

The main body of the report should provide the flow of the argument; those items essential to the production of the 'case'. In most instances, the report will contain a summary, including tables, diagrams and text, of the results of the useable data collected. It is helpful to provide back-up detail and, for assisting future researchers, to include a copy of the data collection instruments used (interview pro-formas, etc.) and details of the responses in the appendices. The 'summary information' provided in the main report is substantiated by the details in the appendices. Thus, the general principle is that appendices provide supplementary information or details, whilst what is included in the main report is the information and data necessary to the development of the argument – in many cases, to the testing of the hypotheses and the fulfilment of the objectives and aim.

Information on the sources and quantities of data sought and of the data actually collected should be included; often, this is achieved with the use of descriptive statistics.

Presentation of results

Presentation of data, tests and analyses executed, and results obtained may be in sequence or within issues such as the nature of respondents, organisational forms and descriptions and so on. Non-essential details should be included as appendices. Whether the presentation is issue-by-issue or using a data–tests–results format is a matter of choice. The format which presents the information in the way in which the likely readers will assimilate and understand it most easily should be adopted. For any but small studies, the issue-by-issue approach seems to be more effective, followed by a discussion of the results which considers the compatibilities of the results – one with another – and with the contents of the reviews of theory and literature; that may be best as an individual chapter following the presentation of results but preceding the conclusions.

Generally, the methods and tests used for analyses of data are well known but choices of which methods and tests to use must be justified. Quantitative analyses include correlation and regression, averages, and measures of variability, rank hierarchies and measures of association. For such well-known standard tests, it is neither necessary nor desirable to include detailed formulae and descriptions of the tests in the report – a reference to a standard text is appropriate. However, on occasions, more 'obscure' tests will be employed; it is appropriate to adopt the same approach to inclusion of their descriptions as for inclusion of the data obtained; essentials in the report and supplementaries in appendices. Analytic methods for qualitative data include discourse analysis, conversation analysis, pattern production and matching, social network diagrams. It is useful to discuss with the supervisor what should appear in the main report and what should be included in appendices.

Discussion of results

Discussion of the results of empirical work requires the results to be evaluated in the context of each other and in conjunction with the theory and literature as well as in respect of the objectives and so on. In this chapter, causations should be examined and explanations weighed up. The initial model of the research variables and relationships may be modified in the light of findings of the study (in which case, further research, perhaps as another project, will be required to test the revised model).

To obtain appropriate detail of communication, consider what information must be included to support and demonstrate the argument and how that information should be provided to communicate the message best. Full details of the results should be provided in the appendices. The results should, as appropriate, be qualified regarding the confidence with which they can be relied upon – if possible using statistical measures but, otherwise, subjectively will be helpful.

Up to this stage of producing the report, the content is impersonal and does not reflect the views of the researcher. The content is facts – theory, findings from previous research, data collected, tests and analyses executed and results obtained. The discussion of results requires the researcher to evaluate the results against each other and against the contents of the theory and literature *and* to deduce explanations for similarities and consistencies and, more obviously, for differences and inconsistencies using what has or has not been done as the basis for the discussion. *If it is not in the report, including the appendices, it cannot feature in the discussion.* Clearly, the principle of including all relevant matter appropriately in the report must be followed. The discussion of the results section may reveal research which has been done but not included in the report. In this case, the report must be amended to include those items which have come to be recognised as important via the discussion.

Conclusions

Once the results have been presented and discussed, it is appropriate to draw conclusions. Conclusions are, in many ways, the most important part of the report – they note the important things which have been discovered through the execution of the research. Thus, conclusions represent what has been discovered, and which can be used by other researchers and by practitioners. A concise chapter is recommended; a maximum of 2000 words is usually appropriate. Each conclusion should be a 'stand-alone' paragraph containing no new material, but firmly founded on the contents of the study. Each hypothesis merits a conclusion; whether it is supported or not, how strongly and why. The same applies to each objective, an explanation of why, and to what degree, the objective has been realised. A conclusion should be drawn about the aim of the research. Generally, only a very small number of findings which are supplementary to the hypotheses and main objectives are adequate to merit conclusions. In research fields other than quantitative research, such as ethnographic studies, conclusions in the forms of hypotheses, objectives and so on for further study are entirely appropriate.

Limitations

An essential requirement of the researcher is to understand the limitations of the research, especially those that apply to the validity of the results and findings/conclusions and to their reliability. The limitations arise out of the choices made by the researcher during the course of the research, although some are imposed by parameters of the study and constraints – most obviously, resource availability. Thus, appreciation of the limitations is an important indicator of the abilities and awareness of the researcher.

Quantification of the validity of quantitative studies is, often, a necessity for statistical testing through determination of the confidence level for statistical significance (and, hence, inference) of test results. However, for qualitative studies, such parameters are more difficult to determine.

Thus, the research report should note all important limitations of the study – in terms of paradigm adopted, data collected, analyses executed, results obtained, conclusions drawn, and so on. Expressing limitations and their sources, may be helpful in two major aspects:

(1) in helping readers appreciate the contribution of the study;
(2) in identifying what might be helpful in future research to extend knowledge – i.e. to inform recommendations for future research.

Recommendations

Research projects may yield conclusions which are sufficient in confidence and content to lead to recommendations for changes in practice. These are called recommendations for implementation and should be noted under an appropriate heading.

The areas in which the aim and objectives have not been met can be suggested as recommendations for further research. Further suggestions may include any potentially significant aspects which the work revealed, but which were outside the scope of the study. Each recommendation should be included in the report as an individual, 'stand-alone' paragraph. Some studies do not produce any recommendations for implementation. The criterion is not to 'force' either conclusions or recommendations – both must be valid *from the research*; if they are not really valid, any 'forced' conclusions or recommendations will devalue the entire research.

Introduction

The final chapter to be written is the introduction. This chapter outlines the topic and the reasons for the study being undertaken, perhaps in the form of problem(s) or research questions to be investigated. Usually, it is helpful to include a brief summary of the background relating to the topic to give an explanation of context and why the study is considered worthwhile.

The chapter should include clear and concise statements of the aim, the objectives and any hypothesis, with sub-hypotheses, and may explain the methodology, with a brief evaluation of why the selected methods have been adopted in comparison with the alternatives. In many research projects, especially where development of an appropriate methodology has been a major factor in the work, it is likely that a separate chapter is devoted to the description and discussion of the methodology adopted, and what alternatives were considered and rejected.

If possible, it is helpful to include a diagram or model of the main variables – independent, intervening and dependent – and their anticipated or hypothesised relationship. The model is derived from the aim and objectives of the study and depicts the relationships in the hypothesis or hypotheses. The model may, of course, be amended by the results of the research, and a revised version can be shown and discussed in a later chapter in the report. The introduction should outline and introduce the total study – what is to be studied, why and how. It is *not* a summary of the work but may note the contents of chapters which follow. Hence, the logic of writing it last of all the report's chapters.

Remainder of the report

References are items of text taken or derived directly from other documents which have been quoted or paraphrased in the report. They may quote precise extracts from texts; ensure that quotations and so on are not too lengthy and are obtained from a variety of sources to provide a comprehensive and balanced view. A bibliography is a list of texts which have been used in the research as general 'background' reading; those texts will not have been cited in the report, as references will have been. For both reference and bibliography lists, a standard method of referencing must be adopted, and both lists must be complete, accurate and consistent for all the source works used to support the research.

Appendices are useful supplements to the content of the main body of the report. They should be used to provide detail which might be useful for particular purposes, such as follow-up research, but are not essential to the content, understanding and 'flow' of the main report. Commonly, summaries of data collected and 'technical notes' are included as appendices.

Oral presentation

Often, an oral presentation of the work by the researchers is required. Such a presentation allows the researcher to present the work and to answer questions to clarify and expand aspects of the report. On occasions, the research may be presented to an audience at a research meeting or a conference – formality will vary enormously between events. Such an opportunity can be very valuable in clarifying the main aspects of the work and in obtaining feedback and fresh views from people who are not close to the work. Such views should be objective and can be very useful in helping to determine what else should be done and by what means.

In making oral presentations, as with producing the report, it is essential to gear the presentation to the audience – to communicate the important aspects of the research in enough detail that they can be understood, but not in too much detail to bore the audience or to be confusing. It is likely that time for a presentation will be limited and, especially as most presentations are followed by a period in which the audience can question the presenter, only the main aspects of the study can be discussed. For work which is *not* complete, it is good to conclude the presentation with an outline of what the researcher intends to do to complete the study. For 'completed' research, the presentation may end with the recommendations for further work.

As in producing a report, visual aids, projection slides and so on of graphs and diagrams are useful, but projections of major summary messages of the components used to develop the argument are very helpful too. Keep the projected messages clear and concise; one idea on each is an appropriate format. Such visual aids may note the themes of discussion, such as 'Transaction marketing'; 'Relationship marketing'; 'Price = Forecast cost plus mark-up'. These aids help to focus the attention of the audience, to reinforce the topic and to remind. Humour is great if it works but must be included with much care and sensitivity to the likely audience.

Oral presentations should not only inform but, to be used to maximum benefit, should stimulate discussion and, hence, feedback. It is in such situations that new lines of research can emerge (a form of 'brainstorming'). Conducting an oral presentation is likely to require the researcher to 'defend' the work; this usually involves justifying what was done and is to be done. Discussion of research from a wide variety of viewpoints is an essential of research and will demonstrate how robust the work's findings are; how well they withstand criticism and/or how generally they apply. The more rigorously the research has been executed, the more robust the findings will be.

Oral presentations provide an interactive forum in which the research can be tested, extended and improved. Most people intend their criticisms to be helpful and constructive; although it may be hard for the researcher at the time, there is great benefit to be gained from the feedback and input from oral presentations.

Summary

This final chapter has considered the communication exercise of producing and presenting the report of the research, the *dissertation* or *thesis*, and of making oral presentations. Recommendations have been made concerning both the contents and the order of preparation of the report. It is important to allow sufficient time for production, including editing and proof-reading. Get whatever help you can to ensure that the report is written objectively and that it says what you intend. It is through presentations of research projects that their worth is judged.

Index

Note: Figures and tables are indicated by *italic* page numbers